Linac Radiosurgery

Springer
New York
Berlin
Heidelberg
Barcelona
Budapest
Hong Kong
London
Milan
Paris
Santa Clara
Singapore
Tokyo

William A. Friedman, MD
Francis J. Bova, PhD
John M. Buatti, MD
William M. Mendenhall, MD

Departments of Neurosurgery and Radiation Oncology
University of Florida College of Medicine
Gainesville, FL

Linac Radiosurgery
A Practical Guide

Springer

William A. Friedman, MD
Edward Shedd Wells Professor
Program Director and Associate Chairman
Department of Neurosurgery
University of Florida
Gainesville, FL 32610-0385, USA

John M. Buatti, MD
Assistant Professor
Department of Radiation Oncology
University of Florida
Gainesville, FL 32610-0385, USA

Francis J. Bova, PhD
Albert E. and Birdie W. Einstein Professor in Computer-Assisted Stereotactic Neurosurgery
Department of Radiation Oncology
University of Florida
Gainesville, FL 32610-0385, USA

William M. Mendenhall, MD, FACR
Professor
Department of Radiation Oncology
University of Florida
Gainesville, FL 32610-0385, USA

Library of Congress Cataloging-in-Publication Data
Linac radiosurgery : a practical guide / by William A. Friedman . . . [et al.].
p. cm.
Includes bibliographical references and index.
ISBN-13: 978-1-4612-7445-2 e-ISBN-13: 978-1-4612-2176-0
DOI: 10.1007/978-1-4612-2176-0
1. Radiosurgery. 2. Linear accelerators in medicine.
I. Friedman, William A. (William Alan) II. University of Florida.
[DNLM: 1. Radiosurgery—instrumentation. 2. Radiosurgery—methods. WL368 L735 1997]
RD594.15.L55 1997
617'.05—dc21 97-7619

Printed on acid-free paper.

Production coordinated by Chernow Editorial Services, Inc., and managed by Bill Imbornoni; manufacturing supervised by Jeffrey Taub.
Typeset by Best-set Typesetter Ltd., Hong Kong.

9 8 7 6 5 4 3 2 1

Stereotactic surgery began in the early part of this century as the result of teamwork between Victor Horsley, a neurosurgeon, and Robert Clarke, an engineer. In a similar fashion, stereotactic linear accelerator radiosurgery began, and has prospered, as a result of teamwork between neurosurgeons, radiation physicists, and radiation oncologists. Unlike many areas of clinical medicine, radiosurgery requires the input of individuals with special education and interest, from each of these fields, in order to produce an optimal result. We are fortunate to have such a team at the University of Florida. We have each benefited enormously from the insights of the others. The resulting synergy has fueled an immensely enjoyable and stimulating research and development process which is now entering its second decade.

Behind each radiosurgery team member is another team, consisting of an extremely patient wife and children, who have provided the support and understanding without which none of this would be possible. We, the authors, dedicate this book to the best teammates we have ever known, our families.

Contents

1

Introduction

Linac Radiosurgery: A Practical Guide is a unique book designed to provide you with a step-by-step, hands-on guide to linear accelerator (linac) radiosurgical treatment. It presents tried and true methods developed over the past 10 years at the University of Florida. We guide you through decision-making processes involved in radiosurgical treatment delivery, and provide you with practical information needed to assist you in your own radiosurgical practice.

Radiosurgery is a one-time application of a high dose of radiation to a stereotactically defined target volume (see Chapter 11 for information on stereotactic fractionated radiotherapy). It differs markedly from both conventional neurosurgery and conventional external-beam radiotherapy. Conventional neurosurgery seeks to resect the pathologic process physically. The ideal result is complete normalization of the postoperative imaging studies. A patient who is neurologically normal immediately after surgery is not likely to encounter serious delayed neurologic complications. In other words, by the end of the hospital stay, the surgeon knows whether the short-term goals were accomplished and whether the patient suffered any complication. Of course, time is required to determine whether resection of the lesion really was complete.

The goal of conventional radiotherapy is to eradicate or control the local disease process. The treatment is typically delivered in 10 to 60 fractions, involving delivery of small doses of radiation once or twice daily through two to four static treatment fields. Differences in biological sensitivity and repair capacity of normal and pathologic tissue are employed to injure the target lesion selectively. Conventional dose planning usually includes a 2-cm margin of normal brain tissue in the treatment volume to ensure that the pathologic process is entirely encompassed by the relatively inaccurate beam delivery and patient positioning systems. Results are measured in terms of disease-free survival and local control. Irradiated lesions frequently do not undergo substantial involution, and success is equated with the absence of further growth. Radiation-induced complications rarely occur during the actual treatment period; rather, they are more typically manifest months to years after treatment. Patients treated with conventional irradiation techniques require prolonged follow-up (i.e., years) to ascertain continued disease control and absence of delayed complications.

Radiosurgery differs radically from conventional neurosurgery in that it is generally an outpatient treatment with no incision and few acute complications. No acute change occurs in the pathologic process. The radiosurgeon cannot tell immediately after the treatment whether control of the lesion was achieved or whether a complication will occur. As in conventional radiation treatment, the beneficial or deleterious effects of radiosurgery will, in general, take months or years to be manifest.

Radiosurgery also differs in several important respects from conventional radiation treatment. Radiosurgery usually relies on one high-dose treatment as opposed to the multiple low-dose fractions inherent in conventional treatment. Treatment delivery involves multiple stereotactically targeted, arced fields versus a limited number of conventionally simulated, static fields. Therefore, the radiosurgeon relies more on extreme accuracy of radiation delivery than on radiobiological differences in tissue sensitivity and repair capacity. The goal of radiosurgery is to deliver a high dose to the target and a minimal, and; it is hoped, harmless, dose to normal tissue just a few millimeters away.

Radiosurgery, therefore, is a unique hybrid of surgery and radiotherapy. Unlike many procedures in neurosurgery or radiation oncology, a multidisciplinary approach that incorporates the unique knowledge of three separate medical disciplines (i.e., neurosurgery, radiation oncology, and medical physics) is required to produce an optimal result. The neurosurgeons provide expert knowledge of the disease processes most commonly treated with radiosurgery: arteriovenous malformations (AVMs), acoustic schwannomas, meningiomas, metastatic brain tumors, and gliomas. They should have experience and expertise in conventional neurosurgical treatment methods, or they should consult with a neurosurgeon with such expertise as part of the selection process for radiosurgery. Neurosurgeons contribute detailed knowledge of neuroanatomy necessary for delineation of the pathologic lesion as well as detailed knowledge of functional neuroanatomy of the normal brain areas to be spared. This knowledge is critical to radiosurgery treatment planning, which juxtaposes large single-fraction target doses next to normally functioning brain. Neurosurgeons engaged in radiosurgery should be trained in standard stereotactic neurosurgery and should have special training in radiosurgery. Such training can be obtained through intense study of the radiosurgical literature, a dedicated radiosurgical fellowship, frequent visits to centers performing radiosurgery, or a combination of the preceding.

Radiation oncologists also contribute expert knowledge of the various disease processes (particularly malignant lesions) as well as of radiobiology that is critical in selecting appropriate patients for radiosurgery as opposed to conventional radiation treatment. Their participation is crucial in treatment planning and dose selection because knowledge of normal tissue tolerance to radiation and sparing of radiosensitive structures is important in these processes. Radiation oncologists that engage in radiosurgery should be trained in standard neurooncologic treatments. They should also have special training in radiosurgery, obtainable by the previously listed methods.

Medical physicists are crucial to any radiation treatment process because they perform the quality assurance procedures that ensure both mechanical and dosimetric accuracy of radiation delivery. Because radiosurgery involves a level of complexity in both of these areas that is unmatched in any other area of radiation oncology, involvement of a specially trained physicist is essential. Medical physicists have driven the research and development in radiosurgery that has brought this rather old technique into its modern era. Their developments include greatly improved treatment planning software, better integration of images into the treatment-planning process, standards for establishing and verifying precise accuracy of radiation beam delivery, and better quality-assurance procedures.

In the chapters that follow, we will examine a typical radiosurgery treatment day. That day begins with head ring application, and is followed by imaging, treatment planning, dose selection, and, finally, radiosurgery treatment delivery. No special knowledge of stereotactic neurosurgery, radiation oncology, or medical physics is assumed. With practical, how-to-do-it explanations, as well as with accompanying photographs and illustrations, we will cover each phase of treatment. Because assessment of control of the lesion and of complications requires prolonged follow-up, we will discuss our recommended schedule of follow-up for commonly treated abnormalities. We will conclude by summarizing our own experience with radiosurgery treatment for AVMs as well as for benign and malignant tumors in terms of both disease control and complications. Finally, Chapter 11 will introduce fractionated stereotactic radiotherapy, which applies the principles of stereotactic radiosurgery to fractionated radiotherapy, and touches on the future of stereotactically guided radiation techniques.

2

Ring Application

Current stereotactic radiosurgical methodology requires attachment of a stereotactic head ring. The rigidly attached ring allows us to acquire spatially accurate information from angiography, computed tomography (CT), and magnetic resonance imaging (MRI). The images obtained with this ring establish fixed relationships between the ring and the target lesion that are later translated during treatment planning so that the treatment target is accurately placed at the precise isocenter of the radiation delivery device. Because the stereotactic head ring is bolted to the treatment delivery device, it also immobilizes the patient during treatment. This chapter will discuss in detail the equipment and techniques used in the first part of the radiosurgical treatment day: ring application.

Equipment

At the University of Florida, a modified Brown–Roberts–Wells (BRW) head ring is used. Most linac radiosurgical centers use some variation of this equipment. The basic head ring (Fig. 2.1) is a stainless steel ring, 29 cm in interior diameter and 31 cm in exterior diameter. It contains eight slots (communicating exterior to interior) through which the post holders (or post drives) are inserted. The front of the ring is stamped "ANTERIOR" and has two angular protrusions. The post holders are usually inserted into the two most anterior and the two most posterior of the eight possible slots. The upper surface of the ring also contains three holes that accept the so-called feet of the angiography, CT, or MRI stereotactic localizers (Fig. 2.1). The undersurface of the head ring contains two screw receptacles, which allow the ring to be rigidly attached to the CT couch for scanning or, more importantly, to the machine for radiation delivery (Fig. 2.2). In the ring-manufacturing process, the precise location of these holes must be verified to be accurately correlated with the remaining ring geometry, to ensure that the patient is accurately positioned

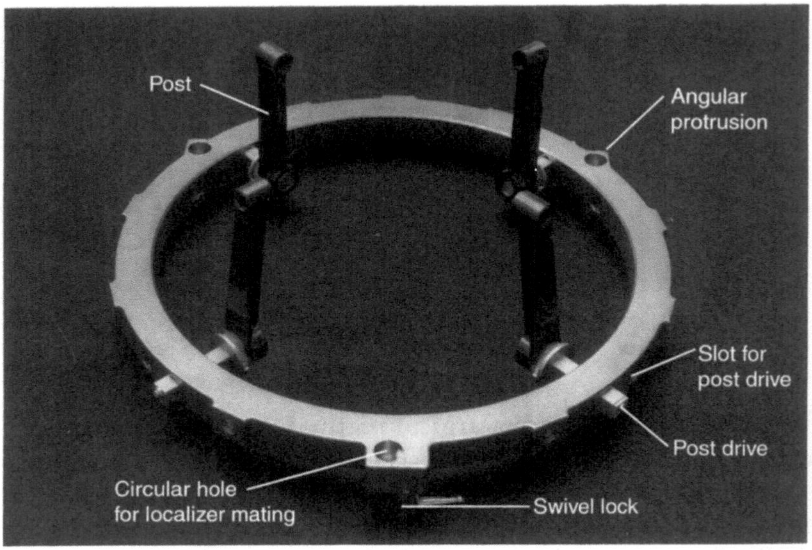

Figure 2.1. The modified BRW head ring as viewed from above. The front of the ring is indicated by the engraved word "ANTERIOR" and by angular protrusions. In addition, the superior surface has three circular holes for mating the localizer feet to the ring. These are controlled by swivel locks on the undersurface of the ring. Posts and slots for post drives are also shown.

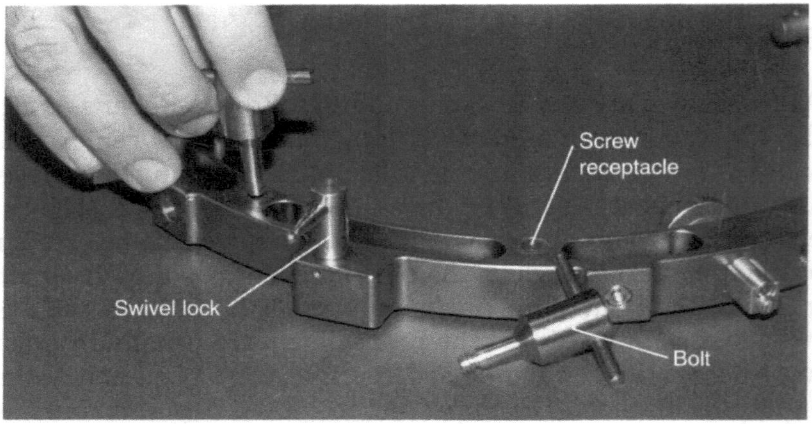

Figure 2.2. The undersurface of the modified BRW head ring. Screw receptacles are shown where the attachment bolts are inserted. The swivel locks control localizer feet attachment sites on the superior surface.

during treatment. The undersurface of the ring also contains swivel locks immediately below the holes on the superior surface that receive the localizer feet. These locks turn to secure the attached feet in position on the head ring (Figs. 2.1 and 2.2).

The posts are composed of carbon fiber. The base of each post contains a stainless steel, circular structure with teeth that mate with a receptacle in the post holder (Fig. 2.3). This allows the post to swivel in a circle about the post drive so that it can be positioned farther anteriorly or posteriorly relative to the patient's head. The top of each post contains a threaded hole through which a pin can be screwed (Fig. 2.3).

The pin is plastic but has a stainless steel tip. It is screwed through the top of the post and through the patient's anesthetized scalp until it touches the outer surface of the skull. Pins are available in short and long sizes (Fig. 2.4). In general, two long and two short pins are used for ring application in an adult. The two short pins are placed anteriorly, and the long pins posteriorly. An adult with an extremely sloping forehead will occasionally also require long pins anteriorly (Fig. 2.5). Four long pins are often used for pediatric head-ring applications. If the head is larger than normal, then four short pins are used to allow clearance around their combined outer circumference for other devices, such as the localizers (see Chapter 3).

This basic head-ring assembly is not MRI-compatible. For cases that require MRI, the equipment has identical geometry but is composed of MRI-compatible materials. The head ring, post drives, and pin tips must consist of MRI-compatible aluminum or titanium. Nonstereotactic MRI scans may be fused instead (with software) to stereotactic CT scans, which obviates the need for MRI compatible stereotactic equipment (see Chapter 5).

Basic Technique

In general, patients are premedicated with 10 mg of oral diazepam (Valium, Roche Laboratories, Nutley, NJ) given approximately 30 minutes before ring application. Premedication is optional. No skin shaving or preparation is required. After the ring is assembled, with post drives and posts approximately positioned for application, the surgeon places the ring roughly in position. The post drives are moved in or out until the post tips rest loosely against the patient's skin (Fig. 2.6). As a general rule, the front pin holes are positioned about 1 in. above the supraorbital ridges and in the midpupillary planes. The back pins are positioned just above the external occipital protuberance, approximately 2 in. from the midline (Fig. 2.7). Ring placement is

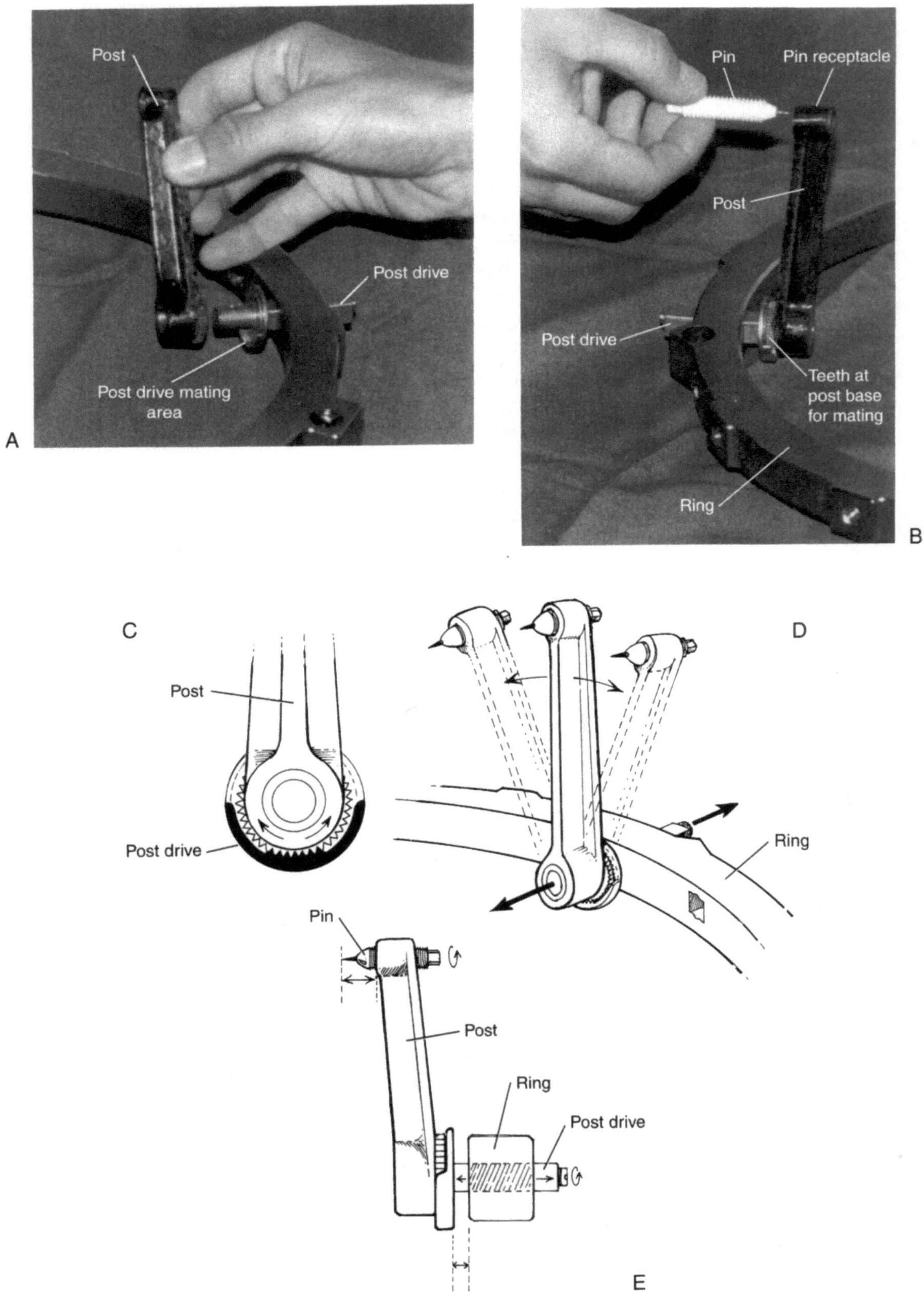
A
Post
Post drive
Post drive mating area
B
Pin
Pin receptacle
Post
Post drive
Teeth at post base for mating
Ring
C
Post
Post drive
D
Ring
Pin
Post
Ring
Post drive
E

Figure 2.3. (A) Lateral view of post and mating area of the post drive before assembly. (B) Close-up lateral view of assembled post and mating area. Post drive, carbon fiber post, pin, pin receptacle, and teeth at the base of the post are indicated. (C) The post base is attached to the post drive, showing the interlocking teeth that allow repositioning of the post (arrows). (D) Adjustments of the post angle on the post drive (dotted lines) and movement of the post drive into or out of the center of the head ring (large arrows). This allows customization of post and subsequent pin position for a given patient's head (i.e., so a bone flap may be avoided and so that post position is appropriate for head size). (E) Illustration highlighting the movement of the post drive through the head ring and screwing of the pin through the post.

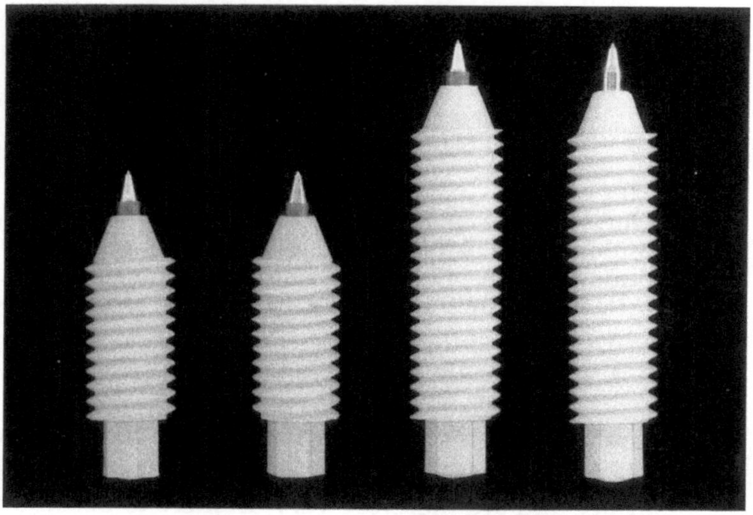

Figure 2.4. Long and short pins.

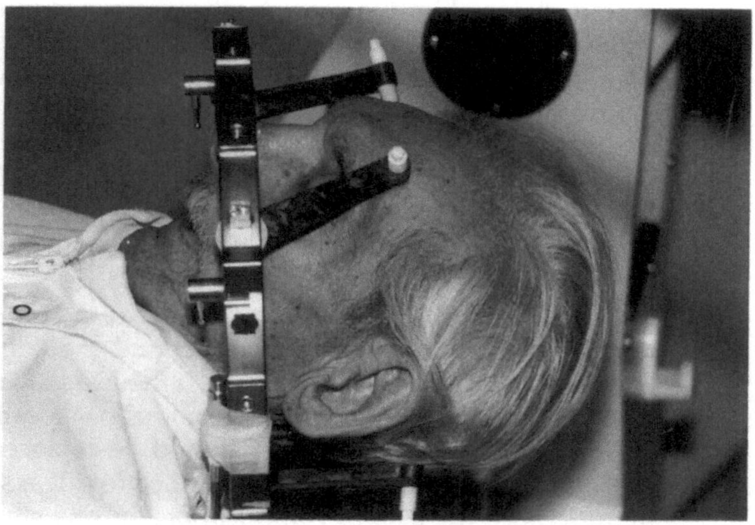

Figure 2.5. A patient with a steeply sloping forehead who required long pins in the anterior posts for ring attachment.

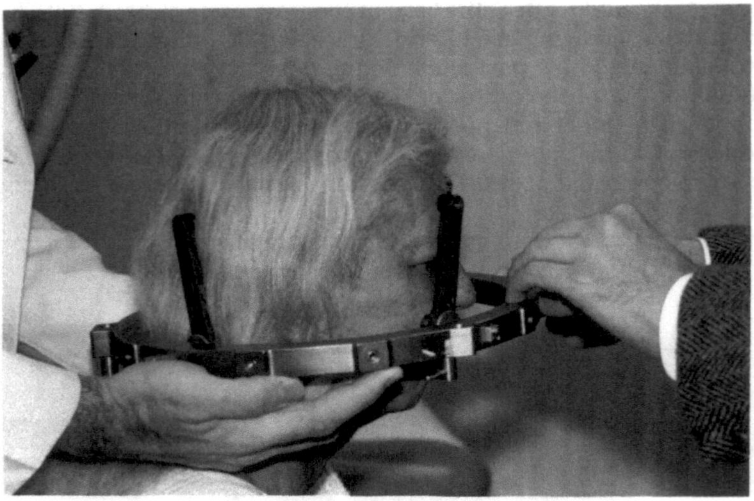

Figure 2.6. The ring is assembled and maneuvered into appropriate position on the patient's head. The assistant loosely stabilizes the ring position from behind the patient while the neurosurgeon adjusts the post positions until they rest gently against the patient's scalp.

usually facilitated by having the patient slightly flex the head. In this position, the pins are usually perpendicular to the skull surface and are therefore very unlikely to become dislodged.

As soon as the head ring is in final position for attachment, an assistant firmly stabilizes the ring from behind the patient while local anesthetic is injected through each of the post tip holes into the underlying skin (Fig. 2.8). A wheal is raised with a solution containing equal parts of 0.5% lidocaine and 0.25% bupivacaine. This solution provides quick onset of anesthetic action as well as long duration.

Approximately 1 minute after anesthetic injection, the pins are inserted into the post holes and screwed through the skin until they rest against the skull (Fig. 2.9). Using the pin wrench, the pins are tightened until the wrench cannot easily be turned using the thumb and first finger only (Fig. 2.10). The CT localizer is then positioned upside down to verify that the inner circumference of the localizer easily passes around the combined outer circumference of the pins (Fig. 2.11). If the head is large, then the combined pin circumference does not allow localizer application. In this case, the long pins should be replaced with short pins. Even four short pins rarely do not provide a small enough circumference for localizer application. In this situation, a few millimeters of one pin are removed with a wire or coping saw. As an alternative for a patient with an extremely large head (i.e.,

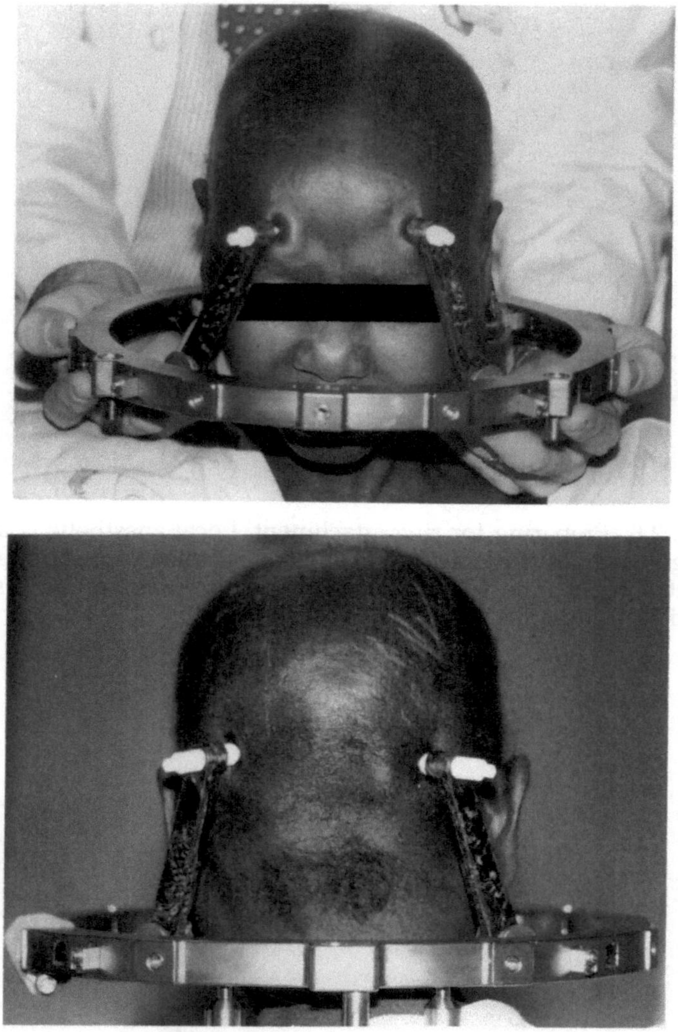

A

B

Figure 2.7. The position of front pins and back pins. (A) Front pins are about 1 in. above the supraorbital ridge and are in the midpupillary line. (B) Back pins are slightly above the level of the external occipital protuberance and approximately 2 in. from the midline.

acromegaly), the post locations can be shifted; the front posts are moved back and the back posts are moved forward, leading to a reduced combined post circumference (Fig. 2.12). (See Number 4 in Pitfalls and Variations in Ring Applications.)

At the conclusion of this procedure, the patient is transferred to a wheelchair and transported to the diagnostic radiology department for the next step (i.e., imaging) in the radiosurgery process.

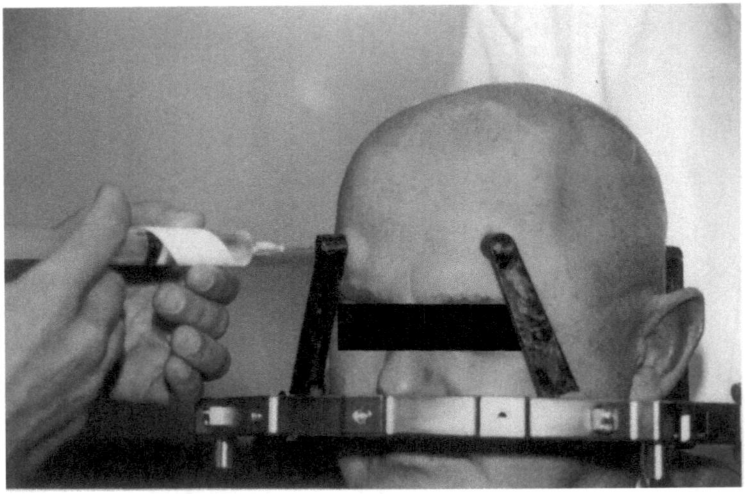

Figure 2.8. Preparing for ring attachment. Local anesthetic is injected through each post-tip hole so that a wheal is created around the site of pin attachment. The subcutaneous tissue through the skull surface should be thoroughly infiltrated with anesthetic.

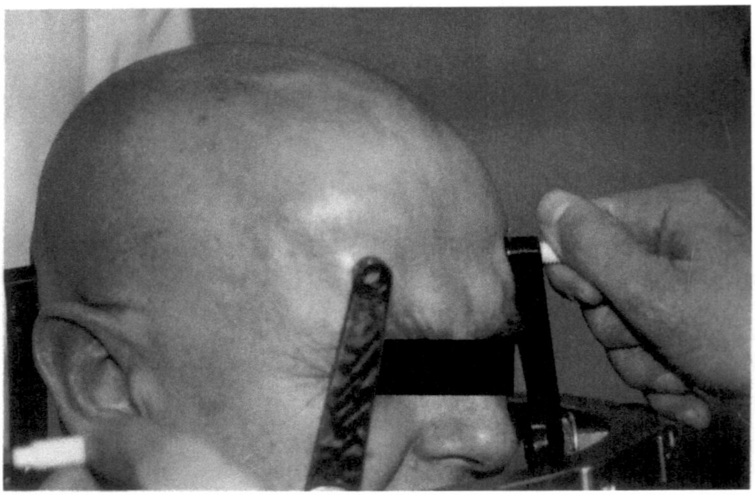

Figure 2.9. Hand-tightening the pins. Each pin should be placed initially by hand so that it pierces the scalp and rests gently against the skull. The patient should not have sharp pain but may have some sensation of pressure. Sharp pain indicates insufficient anesthesia due to inadequate infiltration or due to inadequate time for onset of anesthesia. Sometimes waiting a minute will allow the appropriate numbing; if not, then additional local anesthetic is injected from around the post and aimed at the pin insertion site.

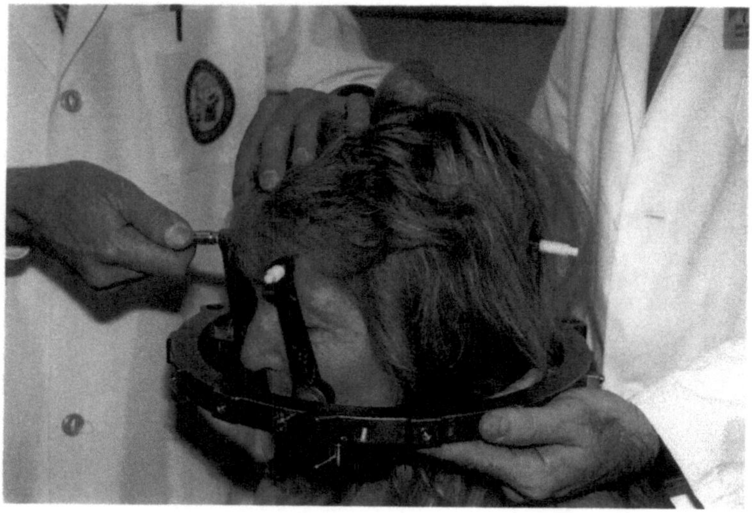

Figure 2.10. The pin wrench is used to obtain final fixation of the ring. The force supplied by twisting the wrench between the thumb and first finger should be adequate for good fixation.

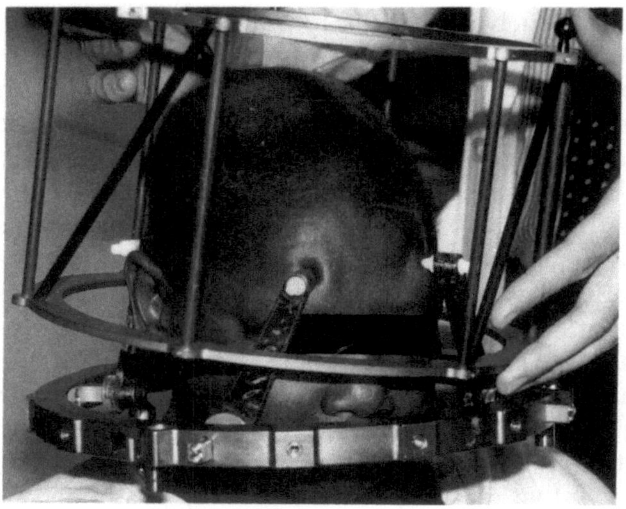

Figure 2.11. The localizer is passed over the top of the head to verify that the pin circumference will allow this passage. Alteration of the pins with a coping saw may occasionally be necessary to allow localizer passage.

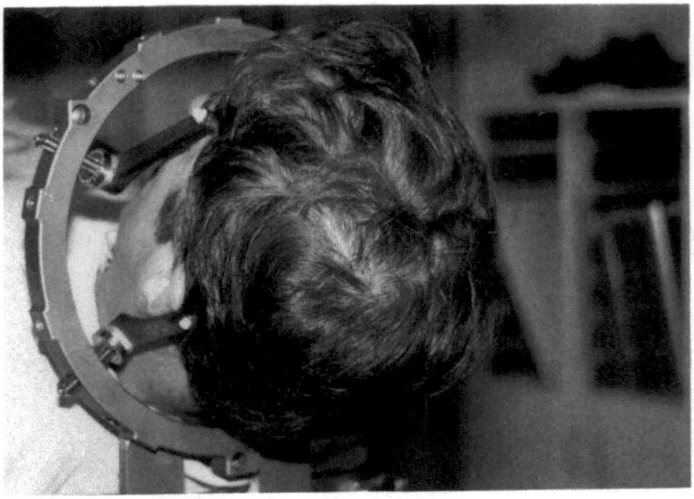

Figure 2.12. An example of ring position in a patient with a large head. Note that short pins are used in both anterior and posterior positions and that the front post has been shifted to the posterior slot and the back post is shifted to the more anterior slot. These maneuvers reduce the pin circumference so that the localizer passes unimpeded over the large head.

Pitfalls and Variations in Ring Application

1. *Failure to image the top of the head.* In an effort to image a low-lying target, such as a jugular foramen schwannoma or posterior fossa AVM, the head ring is sometimes positioned so low that the top of the head is above the top of the angiography, CT, or MRI localizer. The best way to avoid this problem is to have an assistant stabilize the head ring in the tentative application position before local anesthetic injection, and to apply the CT localizer to the head ring (Fig. 2.13). By inspection, one can verify that the top of the localizer is above the top of the head. If it is not, then the ring position is adjusted appropriately before pin placement.

2. *Failure to image the target.* The stainless steel head ring is the foundation for imaging and treatment. It is not, however, compatible with CT or MRI. It is designed to hang from the imaging-compatible posts, below the plane of the target. It is therefore imperative that the ring is appropriately positioned for a given target location. This is rarely a problem unless the target is low (i.e., posterior fossa or upper cervical area). In general, if the ring is below the plane of the external auditory canal, an acoustic neuroma can be adequately imaged. Lower ring applications are

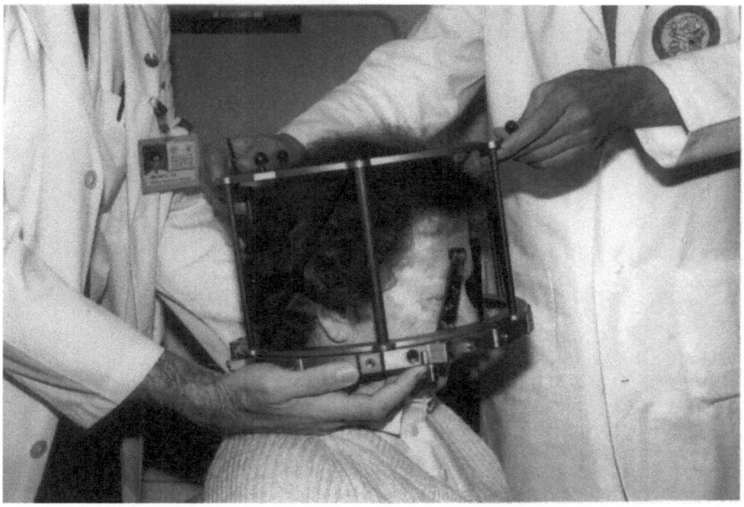

Figure 2.13. The localizer is passed over the top of the head and placed on top of the head ring before injecting local anesthetic to assure that the localizer will include the entire cranium for imaging and treatment planning. Here, the entire cranium is included and the ring position is therefore acceptable. If the localizer did not include the top of the head, then the ring position can be adjusted accordingly.

facilitated by asking the patient to flex the neck and by tilting the ring such that the posterior portion is positioned as low as possible, without having the head ring screws collide with the patient's neck or shoulders (Fig. 2.14). The ring must not be placed too low or the top of the head is not imaged (see Number 1). It is rare, but in a large-headed patient with a very low target, the ring must be applied so low that a small amount of the upper head is not imaged. Although less than ideal, this fortunately results in no appreciable dosimetric error (1.5 to 3%).

3. *Pin artifact.* The imaging planes that contain the pin tips are distorted by artifact (Fig. 2.15). There are two ways to deal with this problem. First, if possible, the ring is applied in such a manner that the target does not fall within the plane of the pins. This is fairly simple to do, but it requires some experience with ring application. Second, so-called artifact-free pins may be used. These disposable pins have aluminum tips, as opposed to stainless steel tips, and generate very little artifact.

4. *The large head.* A patient (generally male) occasionally has a very large head. In such instances it may be impossible to apply the head ring with the post holders in their normal locations. The post holders can be moved to the four more central locations. The front pins are then applied more posteriorly than usual, and

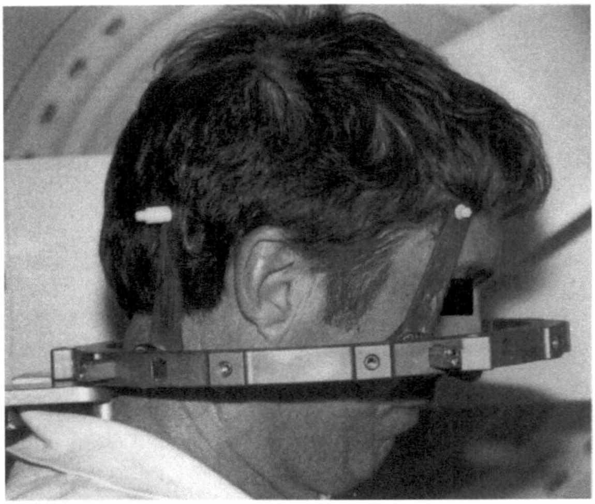

Figure 2.14. A low-lying target can be imaged by getting the patient to flex the neck slightly and tilting the head ring posteriorly. This patient with a jugular foramen schwannoma has a low-lying lesion and a large head that required moving the anterior post back.

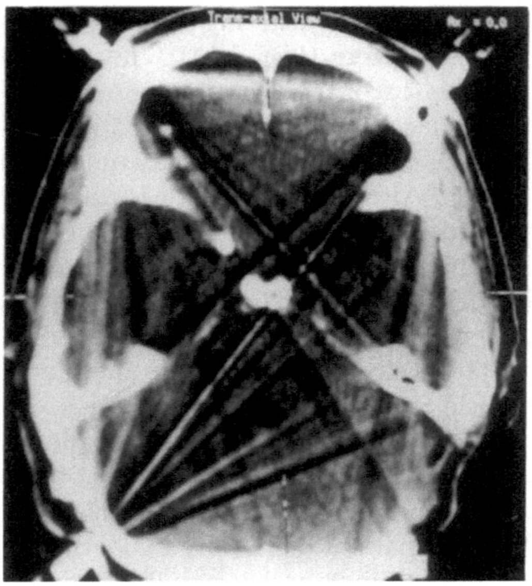

Figure 2.15. Pin artifact is created on CT scan by imaging in the plane of stainless steel pins. Expertise with ring placement and use of artifact-free pins is helpful in minimizing the effect of pin artifact.

the back pins more anteriorly (Fig. 2.12). Using this alternative, we have never encountered a patient in whom a satisfactory ring application could not be obtained.

5. *Prior craniotomy.* Care should be taken to avoid accidentally placing the pins into a burr hole or onto a bone flap from a prior craniotomy. It is occasionally necessary to obtain skull fixation with three pins as opposed to the normal four because a large bone flap interferes. Figure 2.16 illustrates a three-pin application in a patient with a right frontotemporal bone flap. It is also important to avoid puncture of shunt tubing while placing the ring.

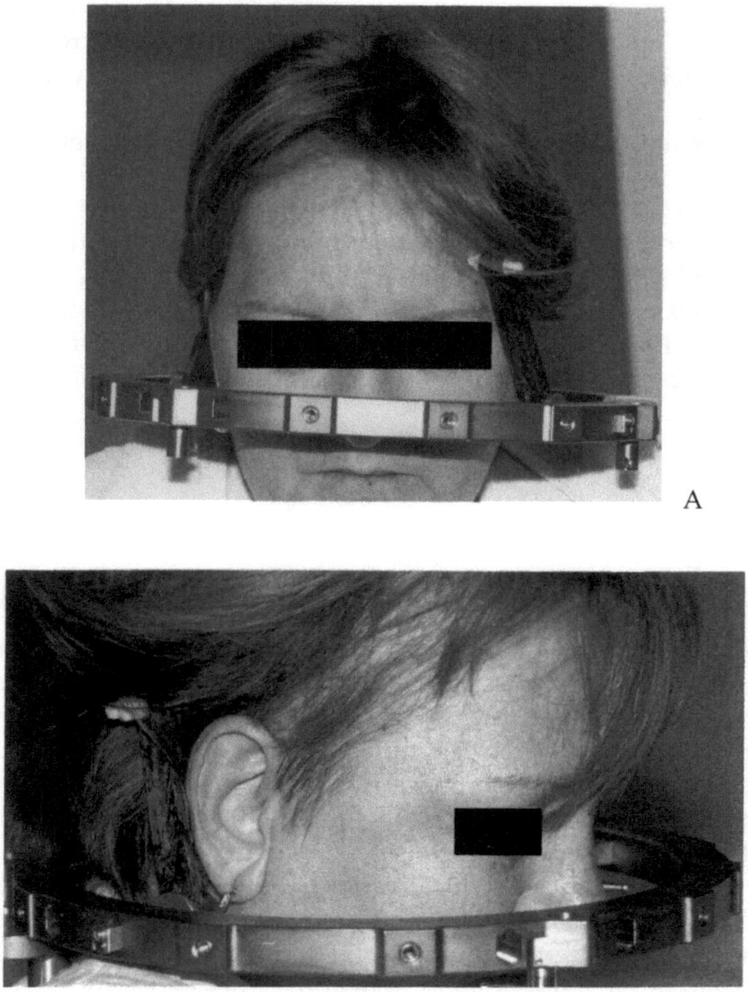

A

B

Figure 2.16. A patient with a three-pin ring application from the front (A) and side (B) to avoid a large frontotemporal bone flap. Burr holes, bone flaps, and shunts must be avoided to prevent complications in ring application.

3

Stereotactic Angiography

Cerebral angiography has long been the "gold standard" for the diagnosis of AVMs of the brain. Indeed, there is currently no better method for precisely defining the arterial supply, venous drainage pattern, and associated vascular anomalies (such as aneurysms) associated with AVMs. In the past, stereotactic angiography was also the only available method for precisely localizing AVMs for radiosurgical treatment. In this chapter, the methodology of stereotactic angiography will be discussed in detail, concluding with a review of some of the inherent drawbacks of angiography as it relates to stereotactic localization. Readers unfamiliar with the definition of *stereotactic coordinates* may wish to read the initial three sections of Chapter 4 first.

Stereotactic Angiographic Principles

Diagnostic angiography typically yields two sets of films that are roughly perpendicular (orthogonal) to one another: an anteroposterior (AP) film and a lateral film. Magnification of the image occurs because the X-ray tube is relatively close to the patient and the X-ray beams diverge. Thus, a diagnostic angiogram displays a magnified image of cerebral vasculature with no fixed references by which the size and location of the lesion can be precisely identified. This is not problematic for diagnostic purposes; however, when images are used for stereotactic localization, issues of magnification and precise location become paramount. To further complicate this scenario, digital angiography is an increasingly popular alternative to the older "cut-film" angiography. Digital angiography introduces additional difficulties regarding spatial integrity. The following methodological issues, therefore, will be discussed: magnification, localization, and digital imaging.

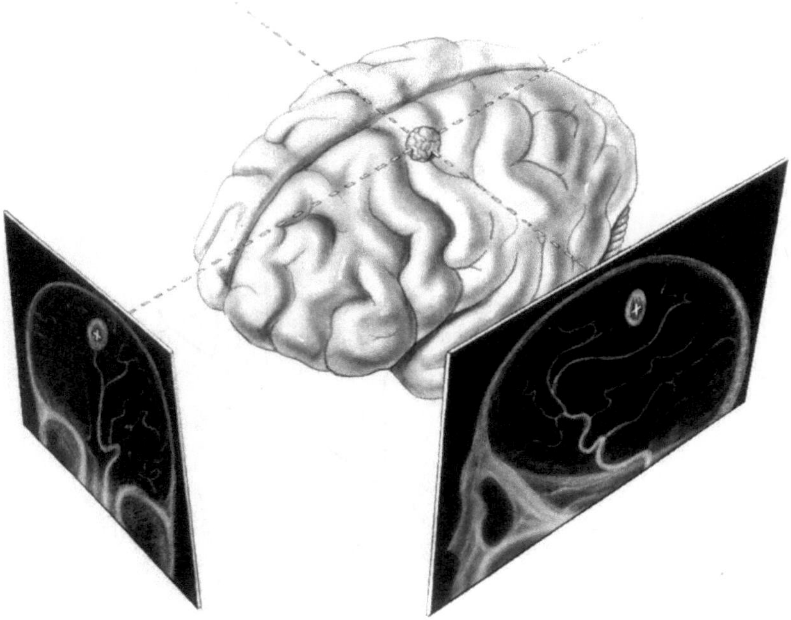

Figure 3.1. A spherical target lesion viewed on orthogonal AP and lateral angiographic films. (With permission from Bova FJ, Friedman WA. "Stereotactic Angiography: An Inadequate Database for Radiosurgery?" *Int J Radiat Oncol Biol Phys* 1991; 20:891–895. Elsevier Science Publishers, New York, NY.)

Magnification and Localization

How can we convert a diagnostic angiogram into a study that yields precise information on the magnification and geometric location of a vascular lesion? The simplest method of solving this problem requires truly orthogonal angiographic images. Figure 3.1 shows a spherical (i.e., three-dimensional) target projected on an AP and lateral angiogram (i.e., two-dimensional). Images obtained from the radiographs are magnified by the ratio of:

$$\frac{\text{X-ray source to film distance}}{\text{X-ray source to target distance}}$$

In Figure 3.2, the X-ray source to film distance is 140 cm and the X-ray source to target distance is 100 cm. This gives a magnification factor of 1.4 (i.e., $\frac{100}{140}$). The geometric coordinates of the center of the spherical target can be measured directly by applying a Cartesian coordinate to the orthogonal films using the center of both planes as the zero reference (Fig. 3.2).

$$\text{AP} = -15.0\,\text{mm}$$
$$\text{Lateral} = 13.0\,\text{mm}$$
$$\text{Vertical} = -9.5\,\text{mm}$$

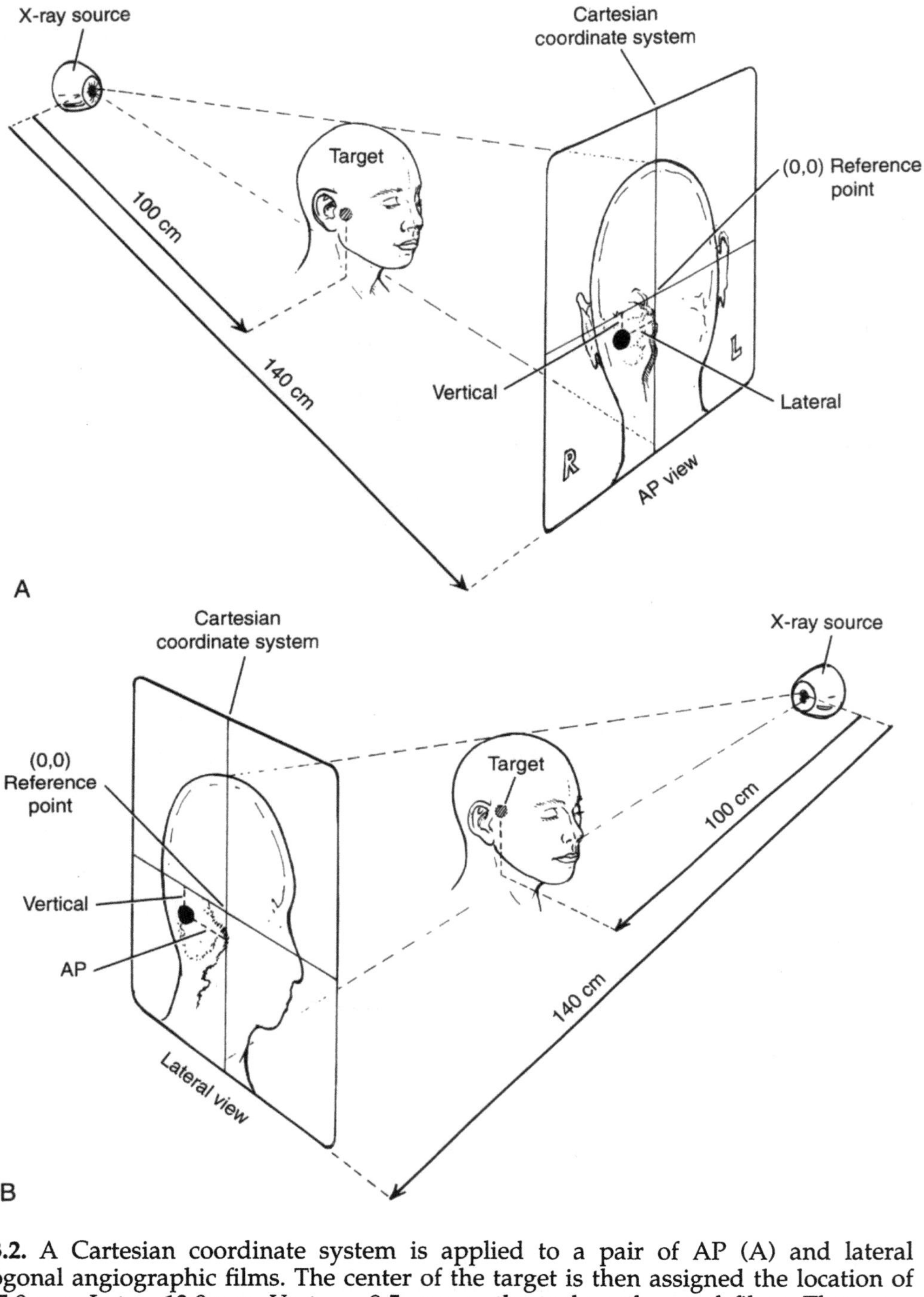

Figure 3.2. A Cartesian coordinate system is applied to a pair of AP (A) and lateral (B) orthogonal angiographic films. The center of the target is then assigned the location of AP = –15.0 mm, Lat. = 13.0 mm, Vert. = –9.5 mm on the truly orthogonal films. These can then be demagnified to give the corrected coordinates of AP = –10.7 mm, Lat. = 9.3 mm, Vert. = –6.8 mm when applying the coordinate system to the patient. Any distance of the lesion away from the center of these orthogonal planes leads to some imperfection in the coordinate representation because the demagnification is based on the center of the films or zero reference. Lesion size can also be calculated from the coordinate system using demagnification based on the center of the system.

After demagnification (dividing each coordinate by 1.4), the corrected coordinates are:

$$\text{AP} = -10.7\,\text{mm}$$
$$\text{Lateral} = 9.3\,\text{mm}$$
$$\text{Vertical} = -6.8\,\text{mm}$$

For this simple case the size of the target can also be computed. Because the target was placed at the center of the image coordinate system, it also has a magnification factor of approximately 1.4. If the sphere projects to a diameter of 29 mm on the AP and lateral radiographs, then the size of the target is approximately 20.7 mm (Fig. 3.2). This approach only works if the lesion is at the center of both orthogonal planes.

While an orthogonal image set allows simple and precise reconstruction, obtaining two truly orthogonal images under clinical conditions is very difficult. Furthermore, it is rarely possible to place the lesion at the precise intersection of the orthogonal planes (as in the preceding example). To solve the problem of magnification and localization in the clinical setting, an alternative and more comprehensive approach is used. One places markers with a known geometric relationship to one another within each image. These markers, called *fiducial points*, are projected onto the angiographic images. They have a known geometric location and a known distance from one another. Hence, they form an internal reference system for the angiogram. This eliminates the need for true orthogonality and allows the use of semi-orthogonal or roughly orthogonal films. Using geometric principles, the size and location of any lesion within the angiographic study can then be computed. The example in Figure 3.3 shows how fiducial markers used in coordination with semi-orthogonal radiographs may be used to determine the size and location of a lesion.

Digital Angiography

Routine radiographic images are analog images, as opposed to digital images. The image contained on the radiograph is continuous. In contrast, a digital image obtained from a digital radiographic unit or that is produced on most computer screens appears the same as an analog image, but, in fact, is made up of small discrete elements each having a defined size and intensity. Each picture element within an image is referred to as a pixel. Routine digital angiographic images usually contain 512 vertical and 512 horizontal picture elements or 262,144 total pixels per image.

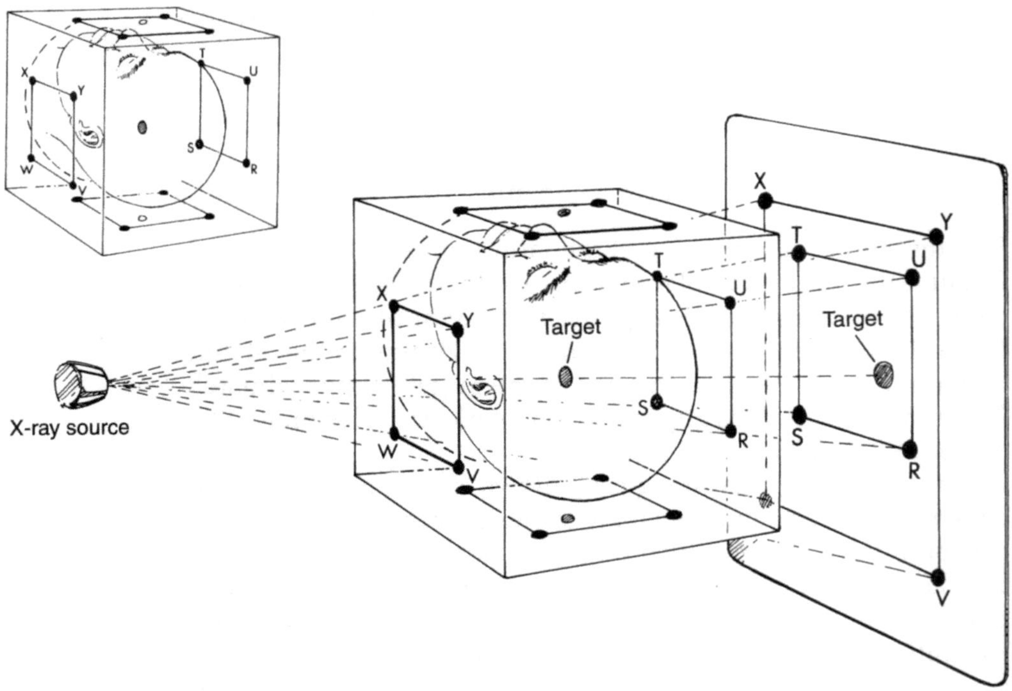

Figure 3.3. An angiographic localizer box is attached to the BRW head ring and contains labeled fiducials in the anterior, posterior, and right and left lateral sides of the box (here represented by the unlabeled points on the anterior and posterior of the box and labeled points on the lateral sides connected to make rectangles). Because the true geometry of these fiducial points is known, the location and size of the target in relation to the head ring can be calculated from the measured distances on the semi-orthogonal angiographic images. The projection of the target on the semi-orthogonal radiograph is shown on the lateral film along with the fiducials. An identical process is used for the AP angiographic film. In practice, a computer performs these calculations.

Advances in digital image acquisition and processing allow greater contrast resolution, more rapid image acquisition sequences, and application of digital enhancement techniques. Digital imaging in angiography involves acquisition of image data through a detector chain that includes an image intensifier, a TV camera, and an analog to digital conversion network. This contrasts with cut film angiography, which involves stacked cut films, a film changer, and a film processor (Fig. 3.4).

Figure 3.4 schematically shows the imaging chains for both routine analog cut film angiography and digital angiography. When the cut film system is replaced with a digital imaging chain, contrast resolution and tissue differentiation are increased at the expense of image distortion. The digital imaging process requires that film be replaced by an image intensifier, which

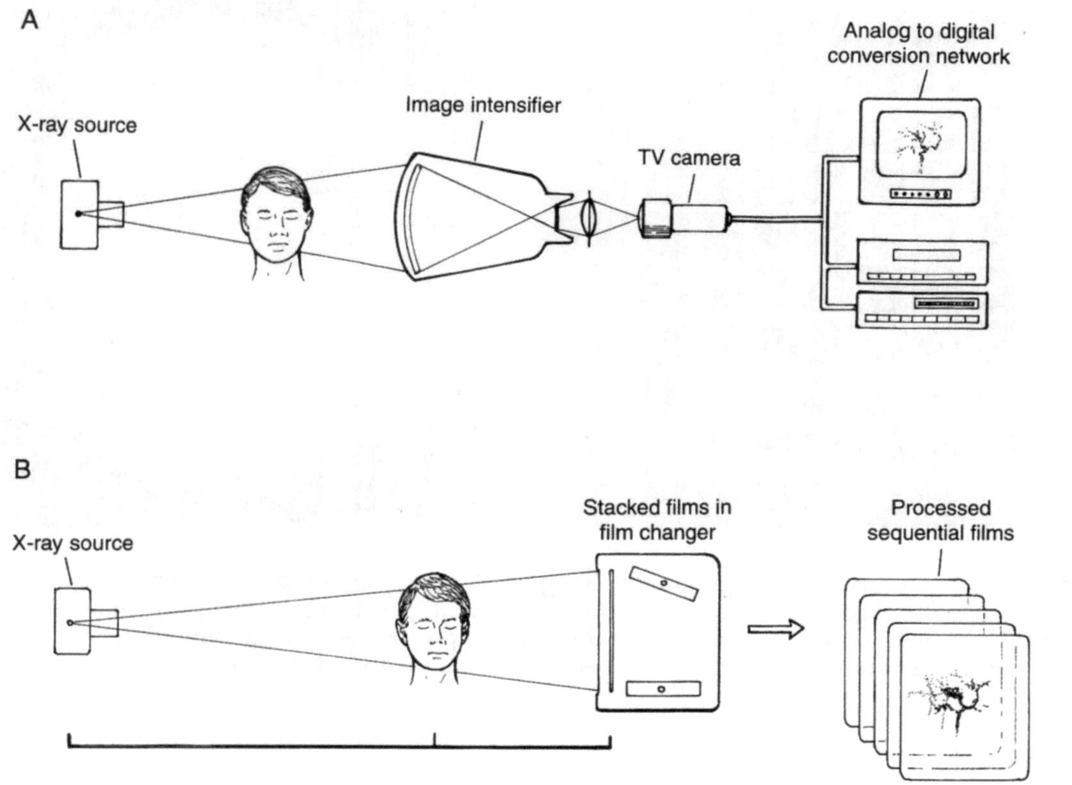

Figure 3.4. (A) The imaging chain for digital angiography begins with an X-ray source interacting with the target but is followed by the interaction of the X-rays with an image intensifier, which activates a TV camera that converts the continuous analog image into digital (i.e., pixel-containing) images. The pixel images can then be manipulated to enhance contrast resolution at various thresholds. (B) The imaging chain for cut film angiography includes an X-ray source that interacts with the target followed by a rapid sequence of plain film angiographic images. The images are continuous (i.e., not pixelated).

contains an input fluorescent screen, an electrostatic lens, and a TV image chain. Each of these digital processing components degrades the geometric accuracy of the final digital image. This distortion is depicted schematically in Figure 3.5. Figure 3.5A is a grid on conventional cut film, and Figure 3.5B is a digital image of the grid. The distortion shown in the digital image increases with distance from the center of the image and is greater than that in the cut film image. This distortion affects the ability to compute the absolute coordinate of a point in space, such as a fiducial point. It also distorts the shape of the projected object.

Image distortion, however, can be measured and a correction can be computed. The distorted grid image when digitally cor-

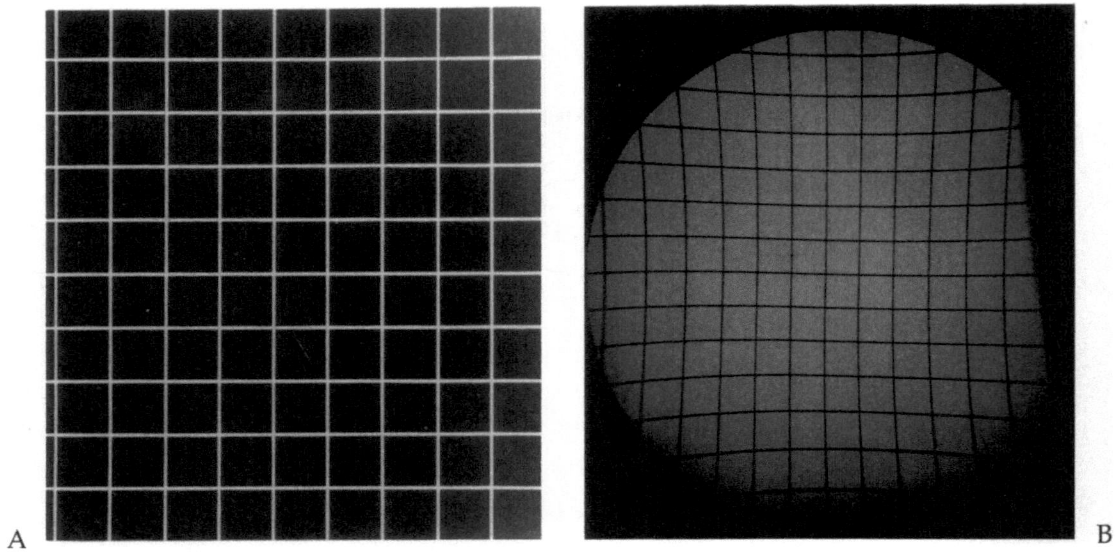

Figure 3.5. Distortions in angiographic imaging. A grid is shown on a conventional cut film (i.e., analog) image (A) and appears undistorted. A digital image of the grid (B) shows the additional distortion that occurs in the digitization process. An image digitally corrected to account for inherent distortion would resemble the undistorted plain film.

rected would appear like the plain film image in Figure 3.5A. In this process, grid points of known geometry in the initial image are automatically detected and the image is remapped to eliminate the distortion.

The bottom line: If digital angiography is used for stereotactic application, then significant spatial inaccuracies are introduced that must be corrected with special computer software.

Paradigm for Stereotactic Angiography Acquisition

After application of the stereotactic head ring, the patient is transported to the angiography suite. The radiologists prep and drape in the normal manner, they do as for any angiogram. A multiple vessel study is typically performed to fully define the vascular abnormality. At the University of Florida, this "survey" is performed using digital imaging to increase speed and reduce the amount of angiographic dye required. When the primary vascular distribution of the lesion is identified, the angiographic localizer (Fig. 3.6) is attached to the head ring. A set of scout films (or fluoroscopic views) verifies that all eight fiducial markers are

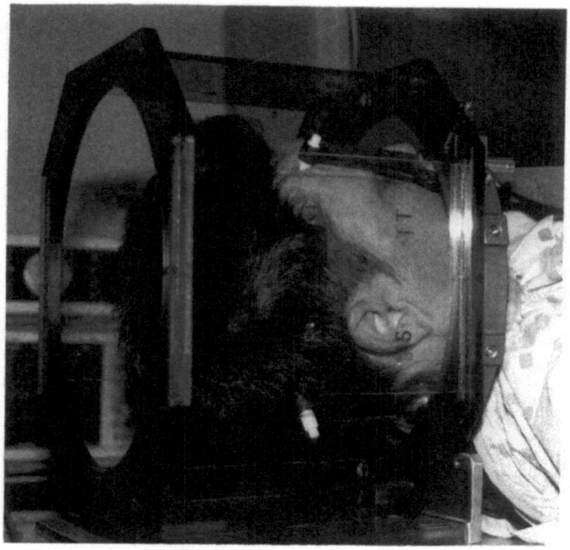

Figure 3.6. A patient ready to undergo angiography with the head ring and angiographic localizer box in position. Note the fiducials labeled by the letters R, S, T, and U on the near (right) side of the localizer box.

seen on the lateral and AP views. A cut film angiographic sequence is then performed (Fig. 3.7). Typically, rapid imaging through the arterial phase is performed, to optimize AVM nidus visualization.

An AP and lateral film that best image the AVM nidus are then selected (Fig. 3.8). The nidus is outlined with a film-marking pencil. These films are taped to a digitizer board (Fig. 3.9). Using a mouselike device, the fiducial points and nidus outline are transferred to the computer screen. Using the previously described geometric methods (Fig. 3.3), the computer determines the AP, lateral, and vertical stereotactic coordinates of the center of the nidus. The demagnified AP and lateral nidus diameters are also computed.

Some radiosurgical computer systems use optical scanners to transfer angiographic cut film images into the computer. The same methodology for solving the magnification and localization issues is then employed.

Pitfalls of Stereotactic Angiography

Angiography is the prime imaging modality for diagnosis and anatomic characterization of cerebral AVMs. It represents the

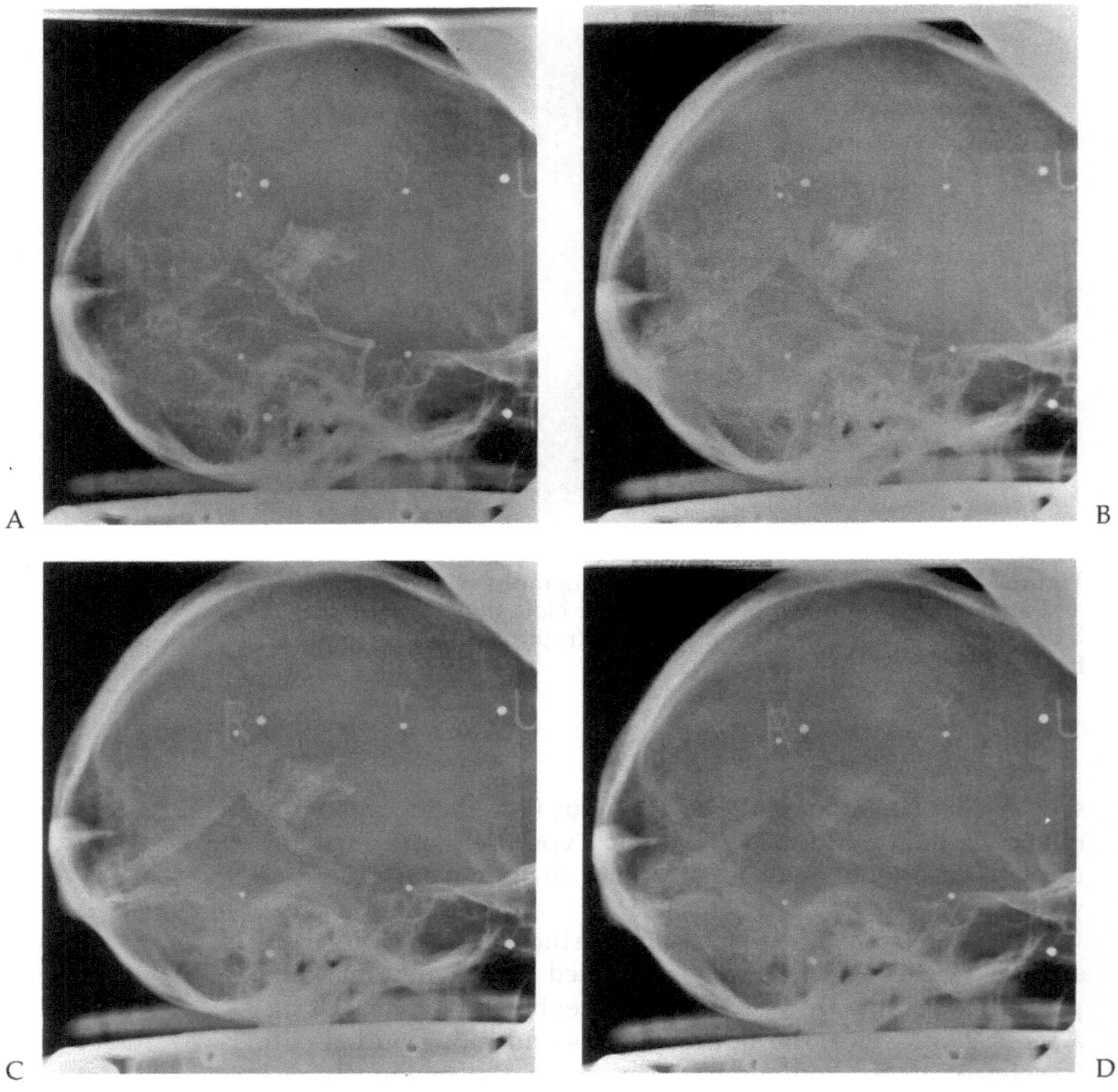

Figure 3.7. Cut film angiographic sequence showing the progression of contrast from the arterial to the venous phase in the AVM nidus. (A–D) Rapid transition from the arterial to venous phase from a lateral film sequence. (E–H) AP view of the same nidus. Note that all eight fiducial markers may be seen on each film.

time-honored means to judge the result of their treatment. To the vascular neurosurgeon, the angiogram provides guiding information required for planning a microsurgical treatment approach. Nidus localization, the origin and approximate trajectory of arterial feeders, and the anatomy of the venous drainage are available from angiographic images. For the neuro-

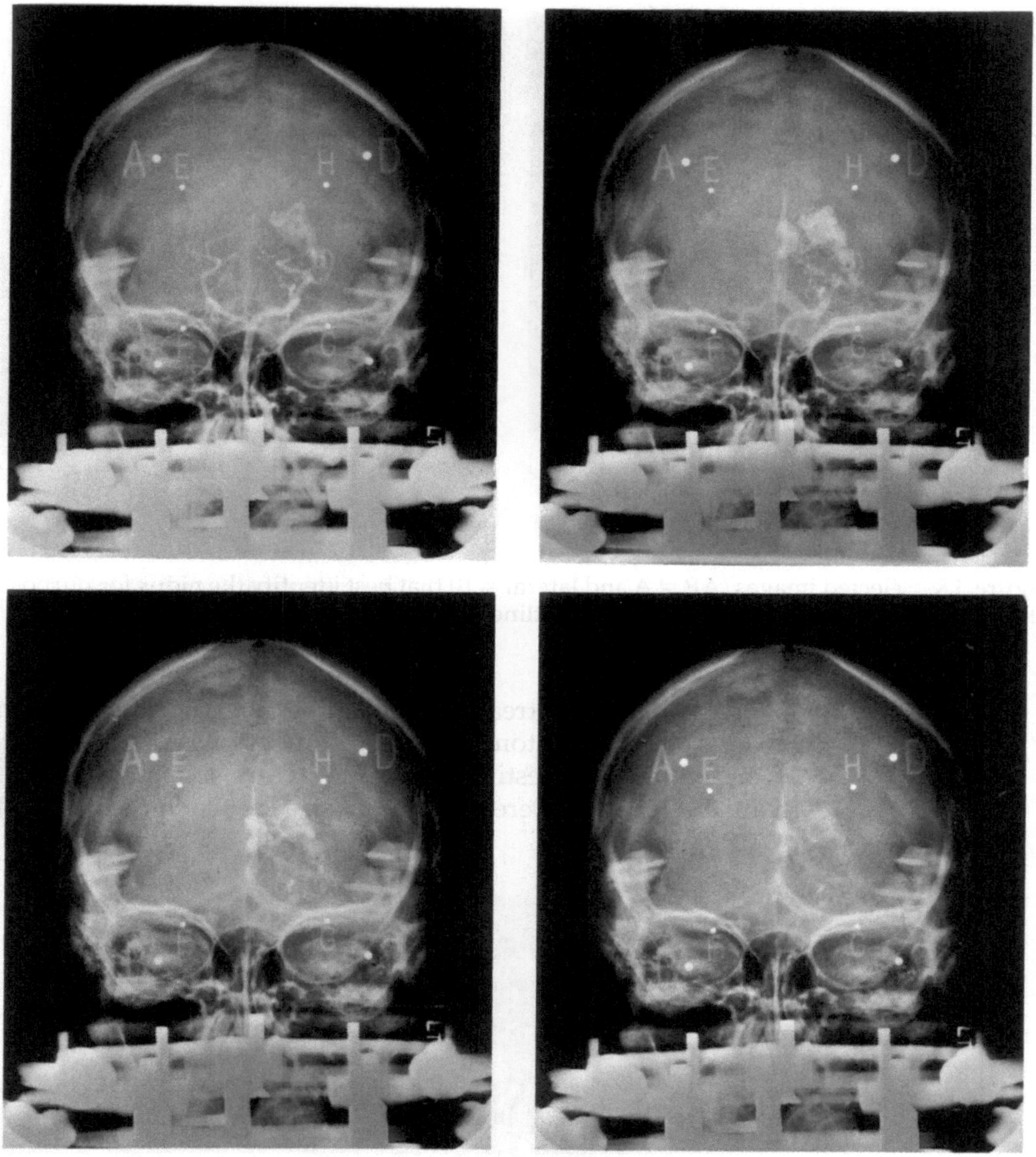

E F G H

Figure 3.7. *Continued*

radiologist attempting embolization of the AVM nidus, the most valuable information in decision making pertains to the nature of the arterial feeders shown on the angiographic study. Disposable terminal arterial feeders that subserve only the malformation are distinguished from in-transit vessels whose occlusion would result in infarction of normal cerebral tissue.

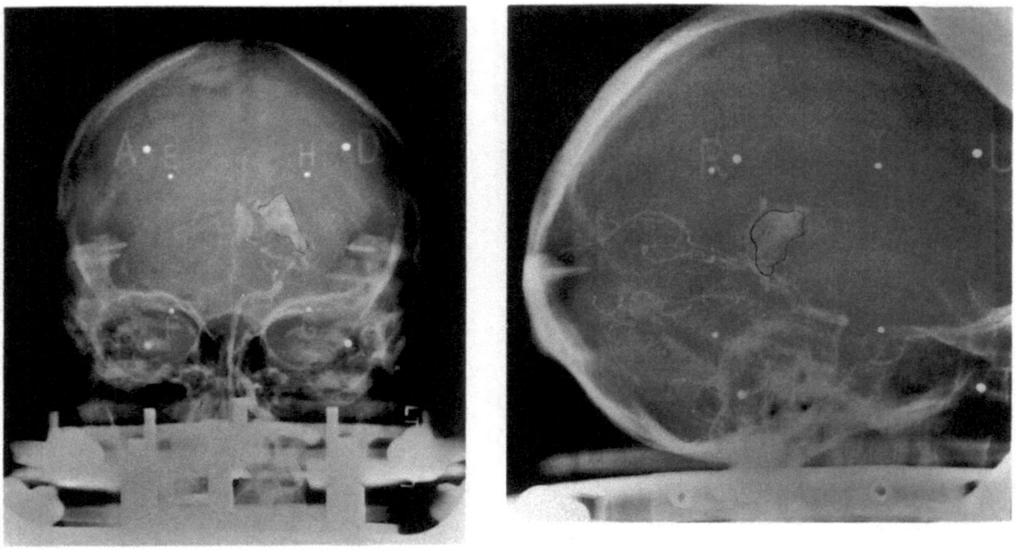

A B

Figure 3.8. Selected images (AP = A and lateral = B) that best identify the nidus for purposes of treatment planning. The nidus is then outlined with a film-marking pencil.

For radiosurgical treatment planning, the two most critical features of AVM anatomy are the tridimensional size and shape of the nidus. Underestimation of the target size may result in treatment failure. Overestimation of size results in the inclusion

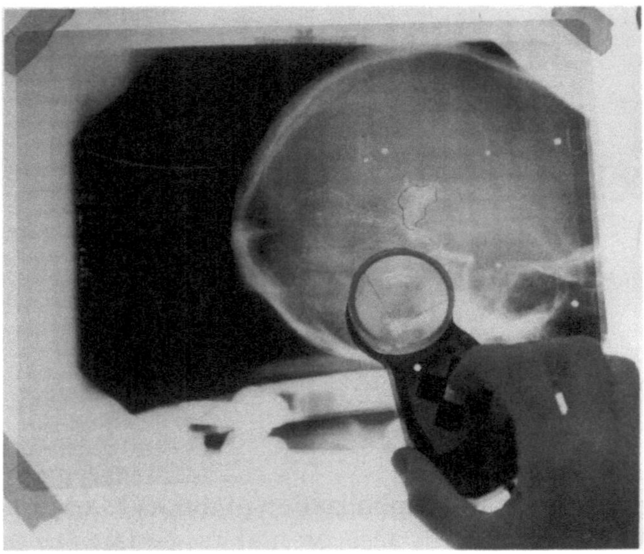

Figure 3.9. The selected images (AP and lateral) are taped to the digitizing board. The outlined AVM nidus and labeled angiographic fiducials are entered into the computer using the mouselike device shown.

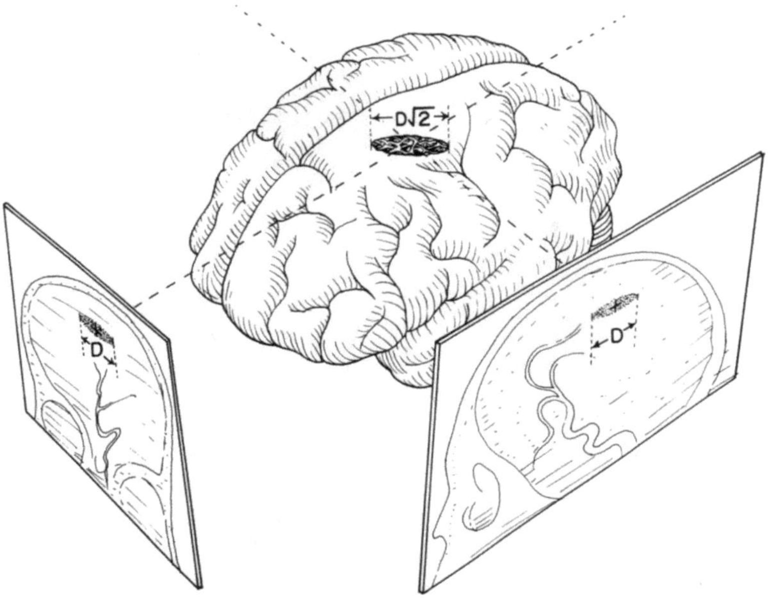

Figure 3.10. Underestimation of an AVM nidus occurs when the nidus axis is not parallel to either AP or lateral angiographic images. Note that the actual size of the lesion is greater than either image would suggest. (With permission from Bova FJ, Friedman WA. "Stereotactic Angiography: An Inadequate Database for Radiosurgery?" *Int J Radiat Oncol Biol Phys* 1991; 20:891–895. Elsevier Science Publishers, New York, NY.)

of normal brain within the treatment volume. Misrepresentation of an irregular target shape may lead to radiation damage of normal brain tissue. When this affects an eloquent area, a neurologic deficit may result.

There are 3 different types of potential errors when an angiogram is used to determine a stereotactic volume. First is *underestimation of nidus volume.* In lesions that have a long axis that is not parallel to any of the projections (Fig. 3.10), the size may be underestimated by angiography. As shown, this error may be as great as 41%. Underestimation of nidus volume on angiography compared with CT treatment volumes occurred in 17% of our cases evaluated for differences.

Second is *overestimation of nidus volume.* Because the angiogram displays a compressed version of the x and z dimensions in the AP view, and, in turn, a compressed version of the y and z dimensions in the lateral view, an overestimation of lesion volume is also possible. Each angiographic view (AP, lateral) represents an aggregate of multiple shapes superimposed on a plane (Fig. 3.11). For nonspherical volumes, the true three-dimensional

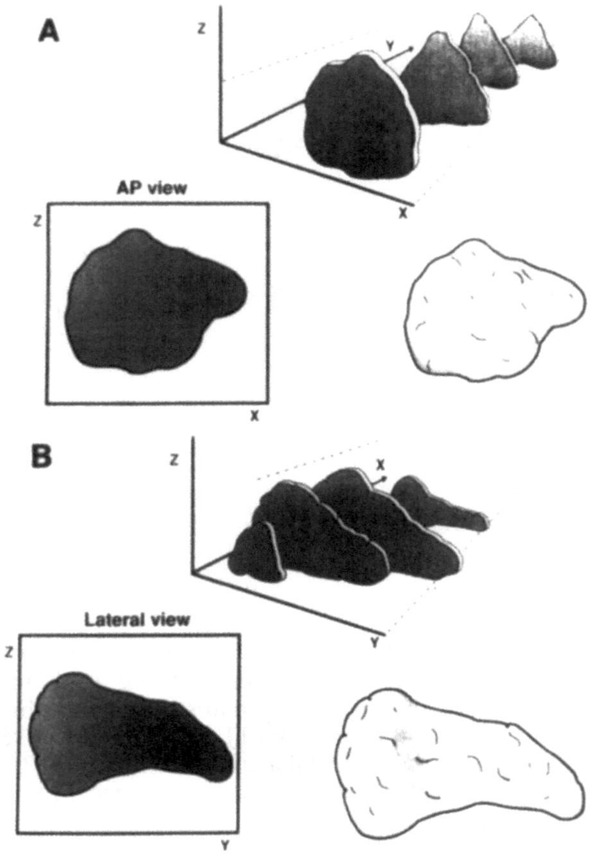

Figure 3.11. Overestimation of the AVM nidus occurs when the image is compressed by the angiogram. (A) The three-dimensional target lesion is represented on the AP view as the compressed x and z dimensions of the lesion. (B) The same lesion is shown from a lateral perspective as a compressed summation of the y and z dimensions on the lateral-view film. Both the lateral and AP views would lead to overestimation of the lesion volume. Treating lesions on the basis of two compressed images in these cases would irradiate normal brain tissue interposed in the concavities of the lesions. (With permission from Spiegelmann R, Friedman WA, Bova FJ. "Limitations of Angiographic Target Localization in Planning Radiosurgical Treatment." *Neurosurgery* 1992; 30:619–624; Williams & Wilkins, Baltimore, MD.)

shape of the target may be poorly represented in one or both views.

Third is *nidus identification (superimposition of different vascular components)*. The high-speed flow inherent to AVMs reduces the angiographic ability to isolate the arterial and venous phases in the area of the malformation. The result is that the projection of the nidus is all-too-frequently intermingled with that of feeding arteries and draining veins (Fig. 3.12). Accurate visualization of

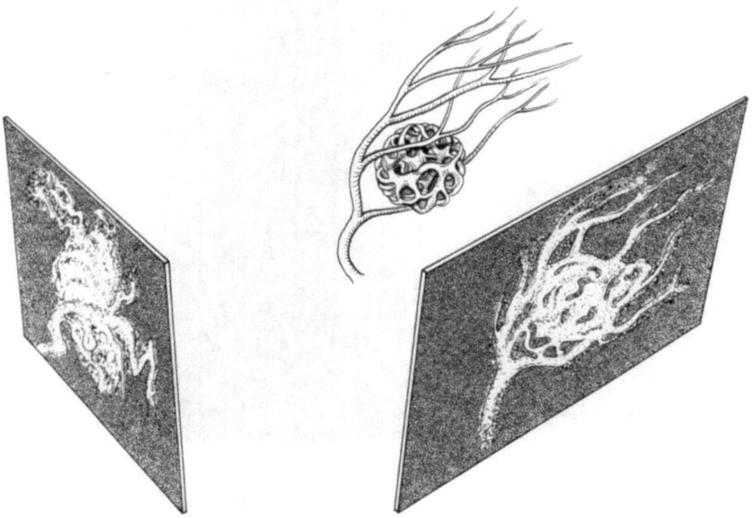

Figure 3.12. It is difficult to identify the nidus on angiography based on the presence of overlying feeding arteries and draining veins. These vessels are compressed in the AP and lateral angiographic images and may confuse interpretation of this spherical nidus geometry. (With permission from Bova FJ, Friedman WA. "Stereotactic Angiography: An Inadequate Database for Radiosurgery?" *Int J Radiat Oncol Biol Phys* 1991; 20:891–895. Elsevier Science Publishers, New York, NY.)

the entire nidus is further compromised by the frequent occurrence of multicompartmented AVMs receiving inflow from bilateral feeders, and/or both carotid and vertebral sources. Furthermore, overlying structures, such as the skull base, may obscure the nidus.

The bottom line: Angiography is a two-dimensional database. AVMs are three-dimensional lesions. Angiography alone frequently fails to indicate the true size and shape of the AVM, leading to errors in dose planning.

For these reasons, the authors strongly recommend supplementing angiography with truly three-dimensional images of the AVM. We usually perform a thin-section CT scan through the area of the AVM nidus. Serial 1-mm cuts are performed while intravenous contrast is infused. This almost always yields a very clear three-dimensional picture of the nidus. Figure 3.13 illustrates a case of an AVM treatment plan imaged by angiography with the estimated nidus circled. The treatment plan based on this angiogram and superimposed on CT is illustrated in Figure 3.13C and demonstrates that a large volume

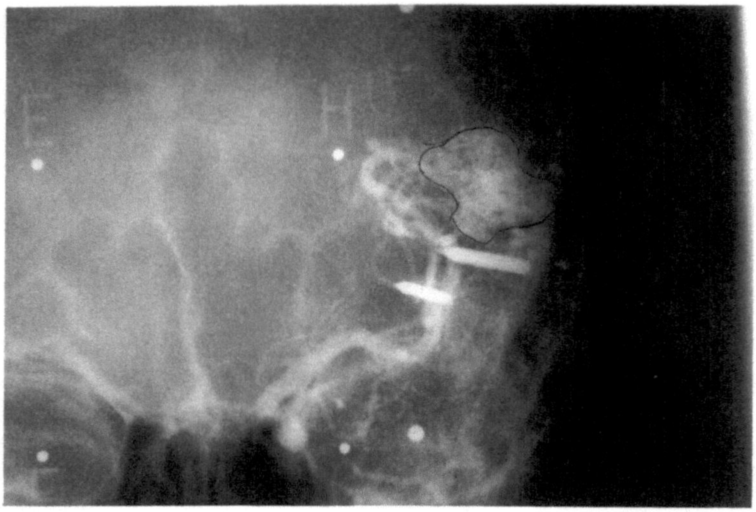

A

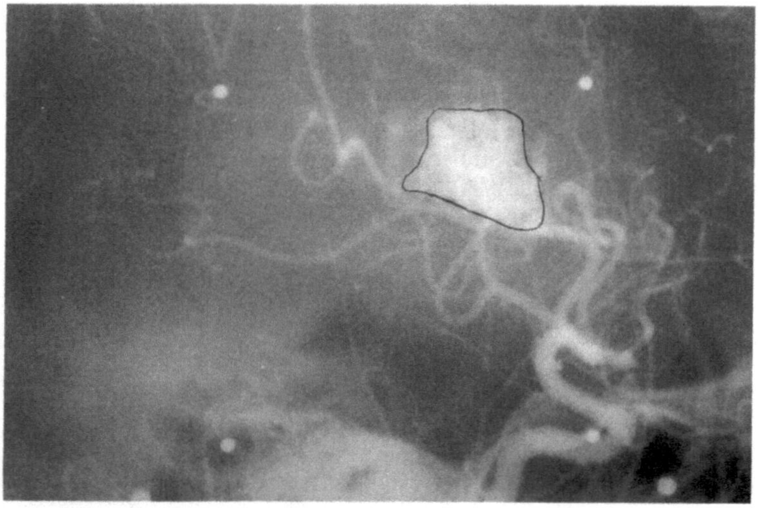

B

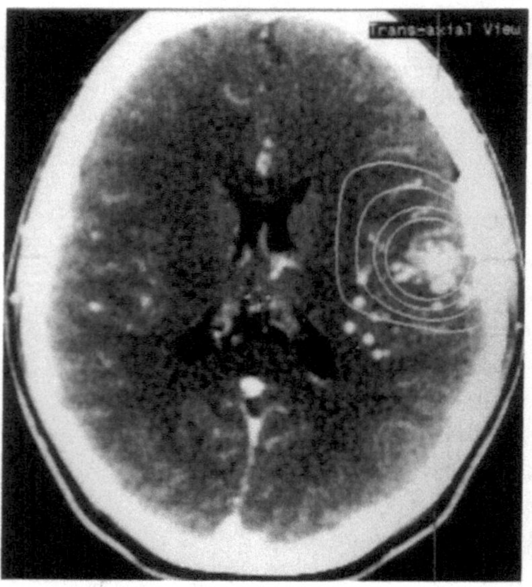

C

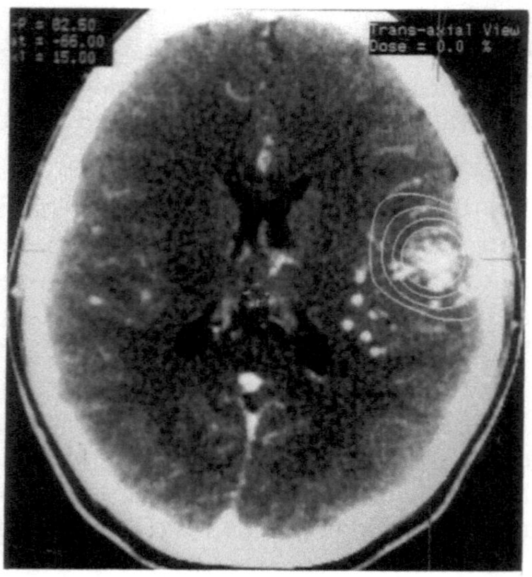

D

of normal brain would be treated by this plan. Figure 3.13D shows the revised plan based on the CT imaging. Others have recommended stereotactic MRI or MRA techniques for the same purpose.

Figure 3.13. Comparison of angiography and CT scan for nidus identification. (A) AP angiographic image used to determine nidus. (B) Lateral angiographic image used to determine nidus. (C) CT image showing an isodose distribution based on the angiographic database. Note that it includes a significant volume of normal brain within the prescription (innermost, 80%) isodose line. The remaining lines shown are the 40%, 20%, and 10% isodose lines, respectively (see Chapter 6 for full discussion). (D) CT image showing the representation of a plan designed from these CT images that enables the radiosurgeon to more appropriately define the nidus and treat a smaller volume of normal brain. (With permission from Blatt DR, Friedman WA, Bova FJ. "Modifications Based on Computed Tomographical Imaging in Planning Radiosurgical Treatment of Arteriovenous Malformations." *Neurosurgery* 1993; 33:588–596, Williams & Wilkins, Baltimore, MD.)

4

Stereotactic Computed Tomography

Background Information: Principles of Stereotactic Surgery

The present state of stereotactic surgery is the consequence of more than 80 years of evolution in experimental neurology, neuroimaging modalities, and, computer technology. The need for a method of exact intracranial localization and reproducible targeting was recognized long ago, but aside from the creation of several craniometric systems that were intended to relate different brain structures to visible or palpable cranial reference points, little progress was made until early in this century.

At that time, while doing basic research on anatomic networks, Sir Victor Horsley became disappointed with his inability to hit the deep cerebellar nuclei using a freehand-directed electrode. Lesions that he produced fell far from the desired target more often than not. He recruited Robert H. Clarke, a young engineer with little previous background in experimental neurology, to help him find "a means of producing lesions of the cerebellar nuclei which should be accurate in position, limited . . . in extent, and involving as little injury as possible to other structures" (Brain 31:45, 1908). Clarke's solution to the problem was simple, original, and enduring. He visualized the brain as a regular geometric body, dividing it with three imaginary intersecting spatial planes, orthogonal to each other: horizontal (axial), frontal (coronal), and sagittal. In this manner, each hemisphere was split into four segments, with each having three deep planar walls and one curved wall corresponding to the brain's surface. Any point within the brain could be specified by measuring its distance along the three intersecting planes (Fig. 4.1).

The place an object occupies in space is determined by its position relative to a given reference point, arbitrarily defined as the intersection of three orthogonal planes. This reference point (defined as zero in all coordinates) establishes a system of axes (Fig. 4.2). The location of any point within the system requires

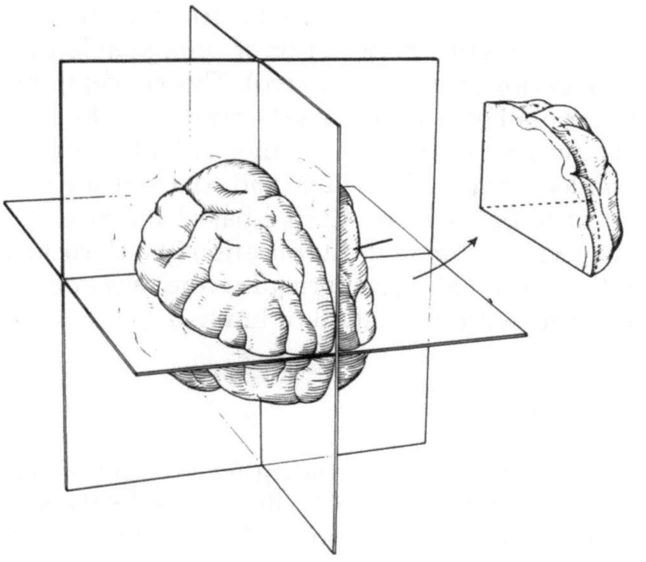

Figure 4.1. R.H. Clark applied a series of three planes to the brain anatomy to define a geometric coordinate system. Note that when this is done, the brain is divided into eight segments, each of which has three deep planar walls and one curved surface.

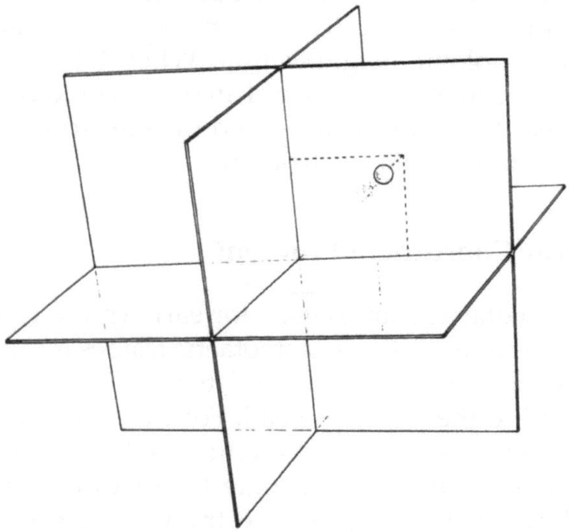

Figure 4.2. Using the intersection of the three planes (i.e., axial, coronal, and sagittal) as the reference point (i.e., 0, 0, 0) allows any position to be defined in reference to those planes. For example, the sphere shown here could be defined as x mm superior to the axial plane, y mm lateral to the sagittal plane, and z mm anterior to the coronal plane such that the point defining its center (x, y, z) is a unique position in space. Positive coordinate numbers designate points superior to the reference axial plane, to the right of the sagittal plane, and anterior to the coronal plane. Negative coordinate numbers designate the opposite sides of each plane, respectively.

the measurement of its distance from zero, in the three planes of space (i.e., z cm anterior or posterior of zero, y cm left or right of zero, x cm superior or inferior of zero). This concept, introduced by the French mathematician, René Descartes, in the seventeenth century, is intrinsic to modern geometry. Its full application to intracranial localization, however, had to await the seminal intervention of Robert Clarke. Clarke based the new concept of stereotactic localization on the application of a Cartesian system of axes to the brain. This system requires the selection of suitable reference points within the skull or brain.

Modern Stereotaxis

Before the revolution in neuroimaging brought about by the introduction of CT in 1973 and MRI in the 1980s, stereotactic localization required conventional X-ray films of the skull, supplemented with gas or dye ventriculography to determine the reference points for application of the stereotactic coordinate system. Stereotactic angiography is a more recent addition. Aside from being time-consuming and painful, exact orthogonal radiologic pairs and precise frame application were critical for avoiding parallax and simplifying cumbersome calculations that account for radiologic magnification. When CT became available, normal and abnormal cerebral anatomy could be displayed in undistorted, scaled axial slices, and the scenario was set for a revolution in stereotactic localization.

The Vertical Coordinate Problem

Although stereotactic frames used for early (pre-CT) stereotaxis had few material restrictions, stereotactic frames for use with CT had to be constructed to obtain undistorted CT images with the frame secured to the patient's head. For the first time, the brain and the geometric system of reference could be visualized together, in linear scale, without parallax or differential magnification. Marking the center of the frame on a CT image and considering, it the zero point was all that was needed; hence, AP and lateral coordinates for any visible intracranial target became directly obtainable.

However, determination of the vertical coordinate for the target was still a problem. Several years elapsed and many short-lived methods were proposed before effective solutions were reached. In 1979, R. A. Brown, a medical student at the University of Utah, applied a simple geometric principle and computerized trigonometric algorithms to derive a solution to the problem. His Lucite prototype, with modifications, became the first stereotactic system entirely designed to interface with CT: the BRW® frame (Brown-Roberts-Wells Radionics; Burlington, Massachusetts). This frame includes three *N*-shaped arrays of carbon fiber rods attached to the stereotactic base ring for CT localization. Each *N* produces three fiducial artifacts (for a total of nine) in any CT slice. The distance of the diagonal rod from the vertical rods allows calculation of the slice height (Fig. 4.3). Determination of the height at three sets of three points (i.e., *N* arrays) defines the spatial orientation of a plane through the frame and the patient's skull, avoiding the need for a fixed relation between the frame and the CT gantry.

In the Leksell® system (Leksell, Elekta; Stockholm, Sweden), introduced in 1980, *N*-shaped aluminum strips embedded into two removable plastic side plates are used to produce fiducial artifacts in each axial image, allowing determination of the vertical coordinate. The Leksell solution, although simpler than the one used for the BRW system in that only elementary calculations are needed (Fig. 4.4), requires that the CT gantry be aligned exactly parallel with the frame's basal plane. This is made possible by magnetic attachment between the frame and the CT couch in this system.

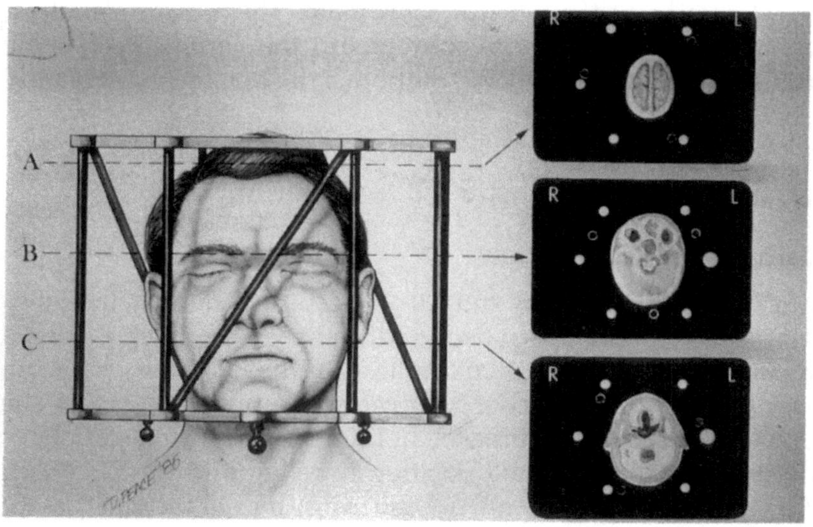

Figure 4.3. The three *N*-shaped arrays of a BRW localizer are used to determine the vertical height of a given axial slice on CT scan. Note that the relationship of the diagonal rods (darkened) in each axial slice to the adjacent rods changes as one moves from slice A to B to C. These differences define the height of the slice in relationship to the head ring.

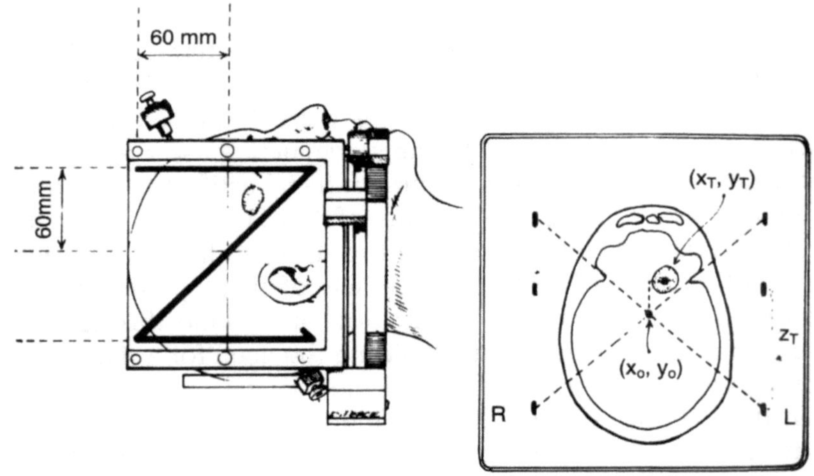

Figure 4.4. The Leksell localizing system uses two *N*-shaped arrays in the side plates. This system requires that the CT gantry is perpendicular to the side plates to accurately determine the slice height.

N-shaped localizing rods have been employed in one of these two approaches in multiple stereotactic systems. The Riechert-Mundinger® frame (Fisher-Leibinger; Dallas, Texas) was adapted to CT stereotaxis with still another geometric variation for calculation of the vertical coordinate. These examples illustrate that there are several solutions to the vertical coordinate problem, although the BRW solution is the one most commonly used today.

Paradigm for Stereotactic CT Acquisition

Basic Scanning Technique

After ring application and, if necessary, stereotactic angiography, the patient is transported to the CT scanner. A special bracket (Fig. 4.5) attaches to the head of the CT table, replacing the usual CT head holder. Screws on the undersurface of the head ring attach to this bracket, holding the head ring stationary and in a fixed, nonrotated position in relation to the CT couch (Fig. 4.6). After attaching the head ring the CT localizer is applied. A CT scanogram is obtained, and the gantry angle is adjusted so that the gantry is exactly parallel to the plane of the head ring, which simplifies calculations, but is not a strict requirement of this procedure (Fig. 4.7). The next step is to indicate where to start and stop the series of axial scans and what slice thickness to use.

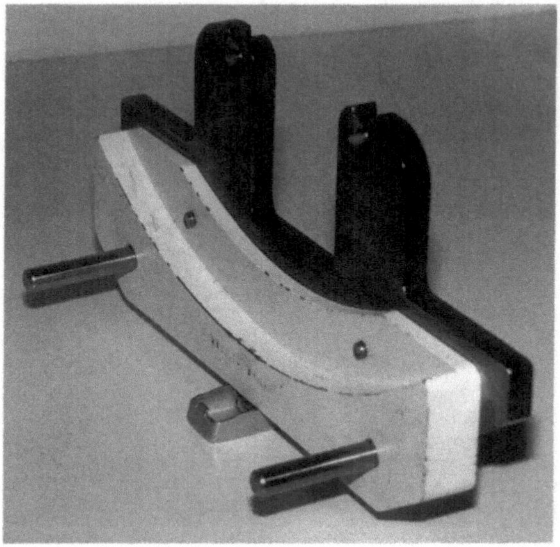

Figure 4.5. The CT bracket is secured to the CT table to assure that the patient is in a fixed and nonrotated position for CT image acquisition. The metal projections (inferior) mate to the CT table, and the two C-shaped receptacles (superior) receive the BRW head ring screws.

We always start the scan at the top of the head (Fig. 4.8). A series of sequential, 5–mm-thick axial scans is obtained down to the area of the target lesion. At that point, sequential slice thickness is changed to 1 mm (Fig. 4.9). After the lesion is believed to be totally imaged, 5-mm slices are resumed (Fig. 4.10). This strategy provides maximum resolution through the target region, but, by using thicker slices in the nontarget brain regions, it reduces the number of scan slices, scanning time, and computer memory requirements.

The diameter of the CT localizer requires a 34.5-cm field of view (roughly equivalent to a small abdominal scanning field) to acquire images of both the cranium and the nine outlying fiducial rods for subsequent CT localization.

Contrast Injection

Patients should not eat or drink anything after midnight the day of the scan in case contrast material causes nausea or other adverse reactions. Nonionic contrast material is preferred because it significantly reduces the incidence of side effects. These side effects are especially undesirable in restrained patients (i.e., those that are clamped to the table). In adults, 100 cc of contrast material is typically used. Scan images are scrutinized during the

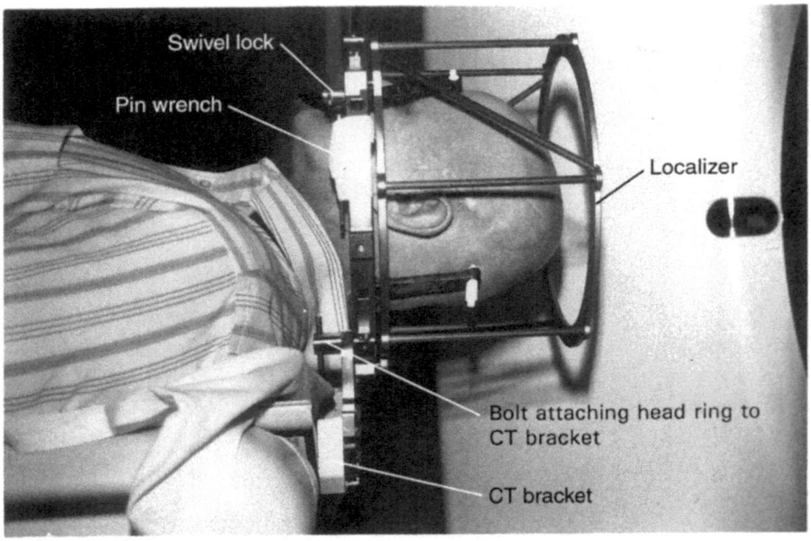

A

B

Figure 4.6. Lateral (A) and superior (B) view of a patient ready for CT image acquisition. After the bracket is attached to the table, the patient is secured to the table bracket by the screws in the undersurface of the BRW ring (see Fig. 2.2) shown on the lateral view (A). Then the localizer is attached, the swivel locks are tightened, and the patient is ready for scanning. Note that the pin wrench is taped to the undersurface of the ring and should remain with the patient at all times in case the frame must be removed in an emergency.

study to make sure the lesion is well visualized for subsequent dose planning. If not, a repeat scan may be done after more contrast material is injected, or after further time delay to allow contrast material to infuse into the lesion.

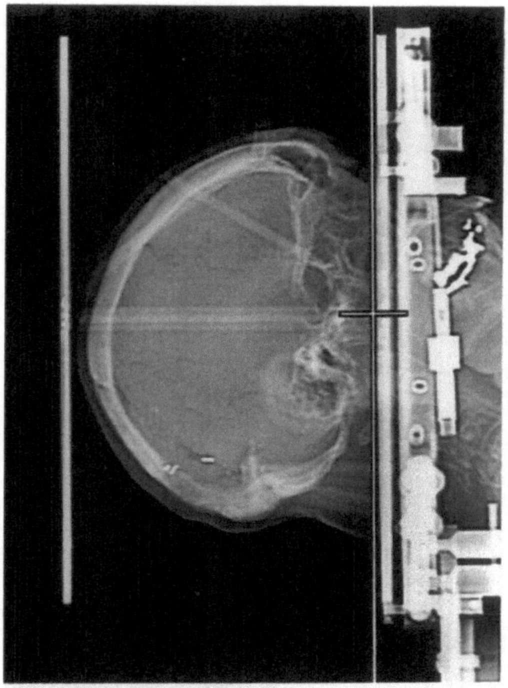

Figure 4.7. CT scanogram in which the gantry angle is represented by the white line at the base of the localizer. The angle of the gantry is made parallel to the localizer ring.

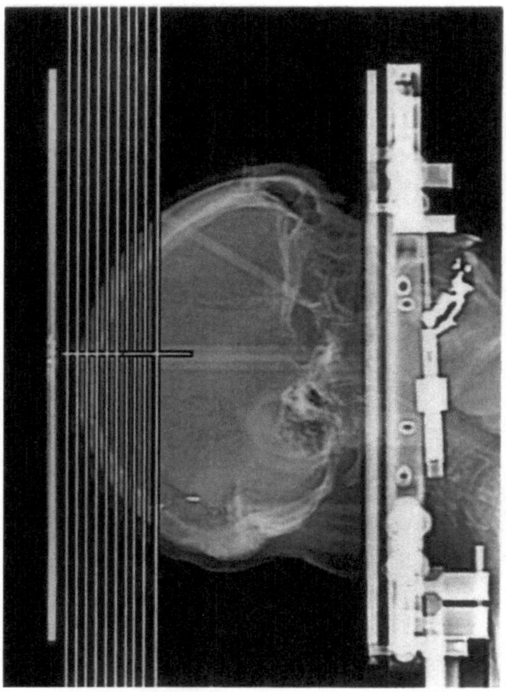

Figure 4.8. CT scanogram showing the start of scanning at the top of the head with 5-mm-thick slices down to the area of interest.

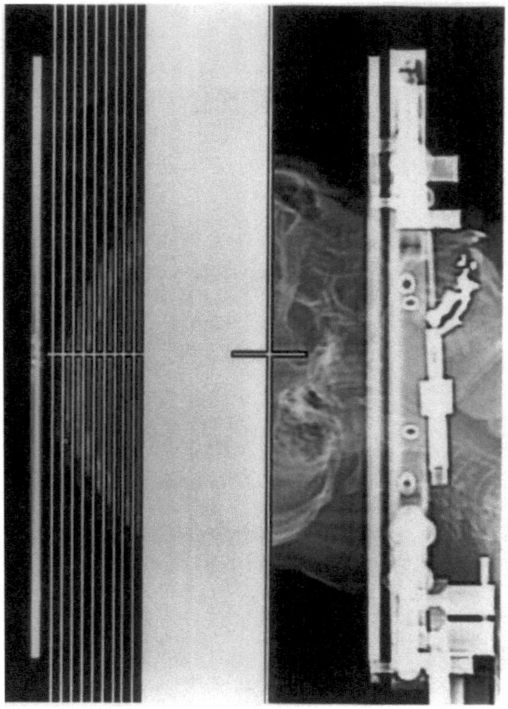

Figure 4.9. CT scanogram showing the initial slices followed by the 1-mm-thick slices (represented by solid white).

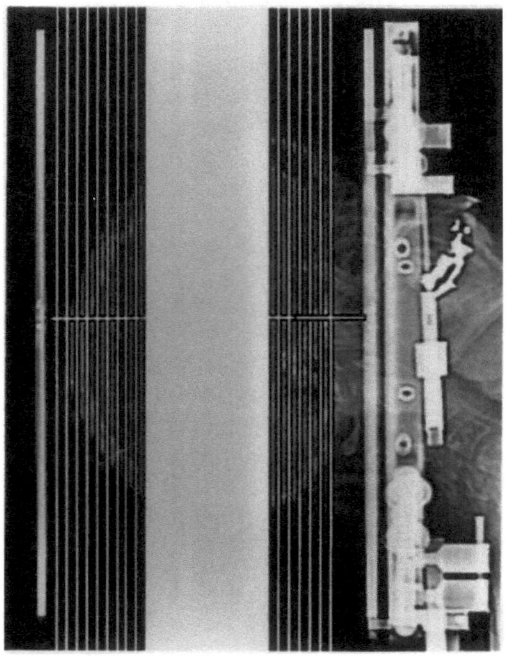

Figure 4.10. Completion of the scanogram with additional 5-mm-thick slices through the remaining brain regions to the localizer base.

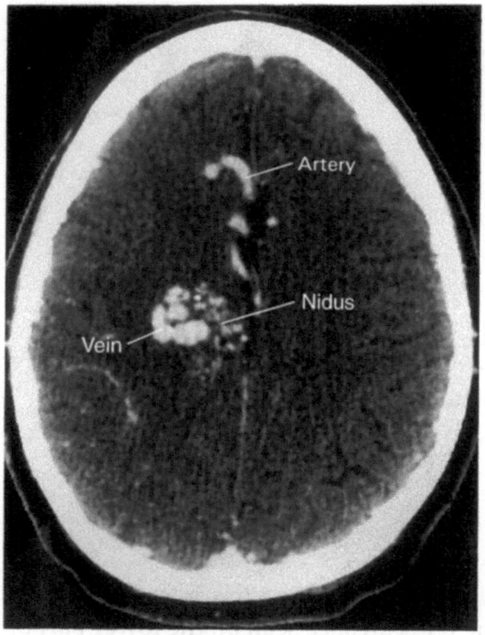

Figure 4.11. CT scan of an AVM using bolus technique for IV contrast, illustrating feeding arteries, draining veins, and nidus. Note the fluffy and compact appearance of the nidus in contrast to the adjacent vein and anterior artery.

When imaging brain tumors, intravenous contrast material is usually injected just before the scanning begins. For AVMs, a special injection technique is used. No contrast material is injected until the top of the target area is reached, when 100 cc of contrast material is rapidly infused (1 cc per second) intravenously as the 1-mm target CT slices are acquired. Scanning while infusing the contrast medium results in a superior image of relevant vascular structures (i.e., feeding arteries, the nidus, and draining veins) composing the AVM (Fig. 4.11).

Image Processing

CT images (80 to 150 CT slices per patient) are next transferred to the dosimetry planning computer. Transfer is accomplished via magnetic tape or via electronic network (i.e., Ethernet) cable. A program in the dosimetry computer automatically identifies the nine fiducial rods that surround each axial image. Using geometric equations, the computer determines the AP, lateral, and vertical position of each point (i.e., pixel) in each CT slice. This information is then replotted in the computer's memory and all CT images are mathematically referenced to the head ring, which remains fixed to the patient's head. Hence, any point seen

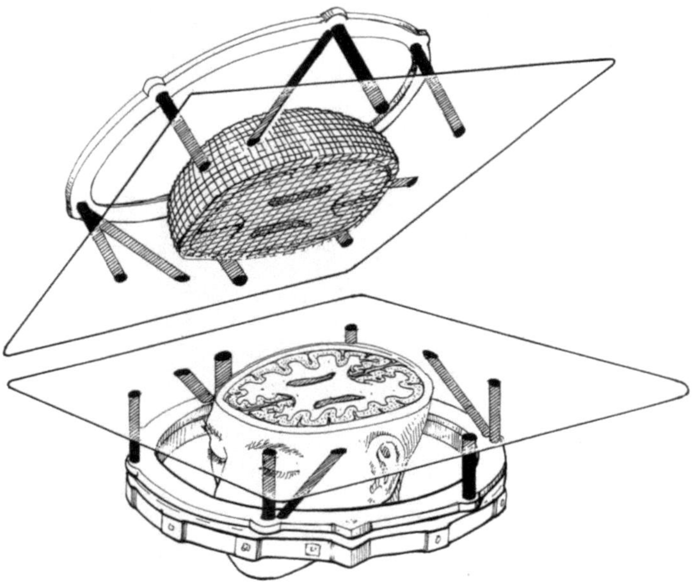

Figure 4.12. The CT scan slice seen on the lower part of the illustration is really a set of discrete pixels as seen in the upper part of the figure. When all the CT scan slices are taken together, they represent a three-dimensional pixelated rendering (above) of the real patient (below) in the figure. Thus, a "virtual" three-dimensional patient exists in the computer that can serve as the template for treatment planning and dose calculation.

on the CT scan image is co-identified as a Cartesian coordinate related to the head ring (Fig. 4.12). Furthermore, because the entire head is scanned and is represented as a conglomeration of unique pixels in the computer, the distance from the scalp to any target point can be mathematically determined from any point along the image. This information is vital for dose calculations because attenuation of each entering radiation beam is proportional to the target depth for that beam. Rapid calculation of dose distribution for hundreds of beams represented by arcs of radiation requires a defined three-dimensional image within the computer. This image is defined during image processing before treatment planning.

Accuracy of CT Scanning

The accuracy of CT scanning depends on the size of the pixels that make up the scan image. The CT screen is typically divided into a 512 by 512 pixel matrix. When using a 34.5 cm field of view, this matrix size corresponds to a pixel dimension of approximately 0.6 mm. Therefore, the AP and lateral accuracy of scanning is approximately 0.6 mm.

The vertical dimension of each pixel, however, is determined by slice thickness. In general, a 1-mm slice is the thinnest that is obtainable with CT scanning. Hence, the vertical accuracy of CT scanning is approximately equal to the slice thickness (e.g., 1 mm in the thinnest slices).

Note that CT scanning, unlike angiography, produces a truly three-dimensional database (Fig. 4.13). As such, it provides an opportunity to understand the true three-dimensional extent of the target lesion. Early in the course of treating AVMs with radiosurgery, it was found that the nidus isocenter and diameter as identified on stereotactic angiography often differed from those seen on stereotactic contrast-enhanced CT. To assess the sources of discrepancy between arteriographic and CT nidus representations, dosimetry (from treatment records stored on optical disc) was reviewed in 81 consecutive cases. Treatment planning using angiography or CT imaging alone revealed that the isocenter locations differed in 44 cases by an average of

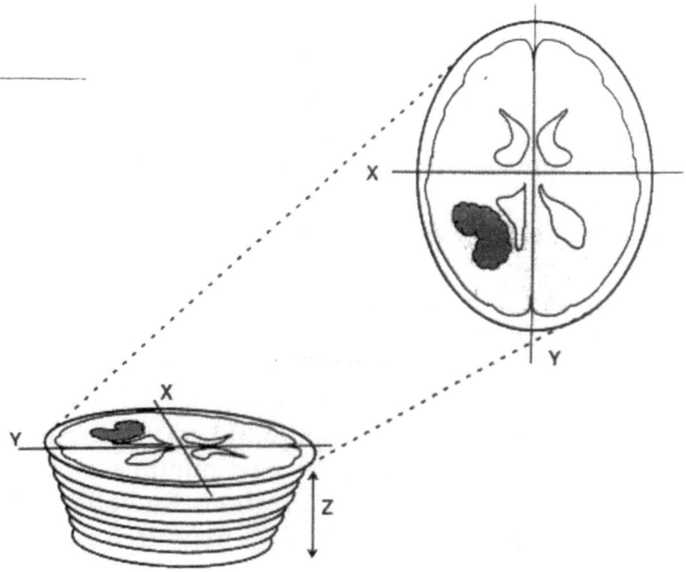

Figure 4.13. A CT scan provides a truly three-dimensional database as opposed to angiography, which can only represent a two-dimensional database. Angiography would represent the three-dimensional target in the AP and lateral dimensions, as shown in Figure 3.11. CT imaging contains a series of axial planes, each of which shows the AP and lateral dimensions. Sequential slices are a decompressed single view and complete the three-dimensional database. Furthermore, the pixels may be reconstructed to show coronal and sagittal images. (With permission from Spiegelmann R, Friedman WA, Bova FJ. "Limitations of Angiographic Target Localization in Planning Radiosurgical Treatment." *Neurosurgery* 1992; 30:619–624; Williams & Wilkins, Baltimore, MD.)

3.6 mm. In addition, the collimator size was changed in 44 cases (Fig. 3.13). Fourteen nidi were larger using CT imaging compared with angiography (average, 2.6 mm), whereas 30 were smaller (average, 4.0 mm). Overall, the angiographic and CT nidus differed in 75% of the cases reviewed. Sources of error in the angiographic nidus determination included overlapping vessels and bony structures, fine filamentous arterioles, and irregular shape. In most cases, we rely predominantly on CT scanning for definitive decision making in the dose planning process for AVMs as well as for other lesions.

Note that CT, unlike MRI, introduces no spatial distortion into the image (see Chapter 5). Stereotactic position of the lesion as seen on CT is, quite reliably, undistorted.

Pitfalls of Stereotactic CT

1. *Pin artifact.* As discussed in the chapter on ring application, the image plane that contains the four metal pin tips is obscured by metal artifact (see Fig. 2.15). Insofar as possible, the ring is applied so that this plane, which usually comprises just one or two of the 1–mm-thick CT slices, is above or below the target. An alternative is aluminum-tipped pins, which are relatively artifact-free. However, because aluminum is softer than stainless steel, aluminum pins require replacement after a limited number of applications.

2. *Other artifacts.* Like the stainless steel pins, other metallic objects in the head can obscure the CT image. For example, aneurysm clips or other metal clips near an AVM or tumor may make imaging difficult (Fig. 4.14). In such instances, an alternative imaging database such as stereotactic angiography may be imperative for target localization. If the metal is MRI-compatible, stereotactic MRI may be another option. In addition, ring application is planned such that the ring angle minimizes metal artifact interference with target visualization. Embolic materials used in endovascular treatment of AVMs may also cause considerable artifact on CT scanning.

In the posterior fossa, artifact generated by the bones of the skull base may reduce the image quality. Most lesions in the posterior fossa are better visualized with stereotactic MRI.

3. *Lesions poorly seen on CT.* Some lesions, predictably, are poorly visualized on CT. For example, intracanalicular acoustic neuromas are rarely well imaged, even with thin-section CT (Fig. 4.15). Jugular foramen schwannomas are very difficult to visualize using CT (Fig. 4.16). Certain AVMs, especially those less than 15 mm in size, may be better visualized on angiography (Fig. 4.17).

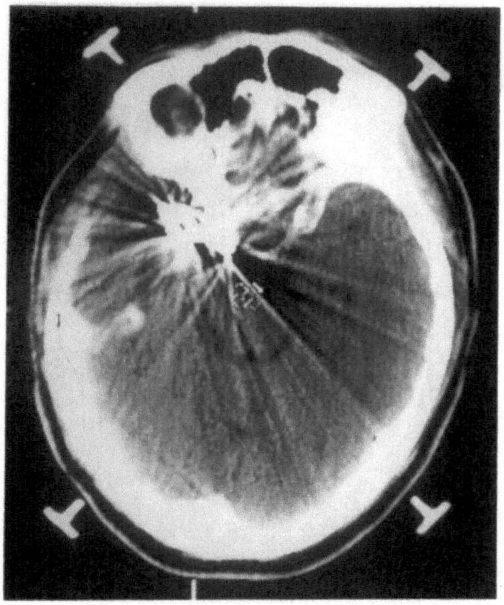

Figure 4.14. CT scan illustrating artifact produced by aneurysm clips that had previously been used for the treatment of this AVM. Such a case may preclude adequate use of CT scanning and may best be treated based on imaging with MRI, if the clips are compatible, or with angiography if it is impossible to obtain a three-dimensional database.

A 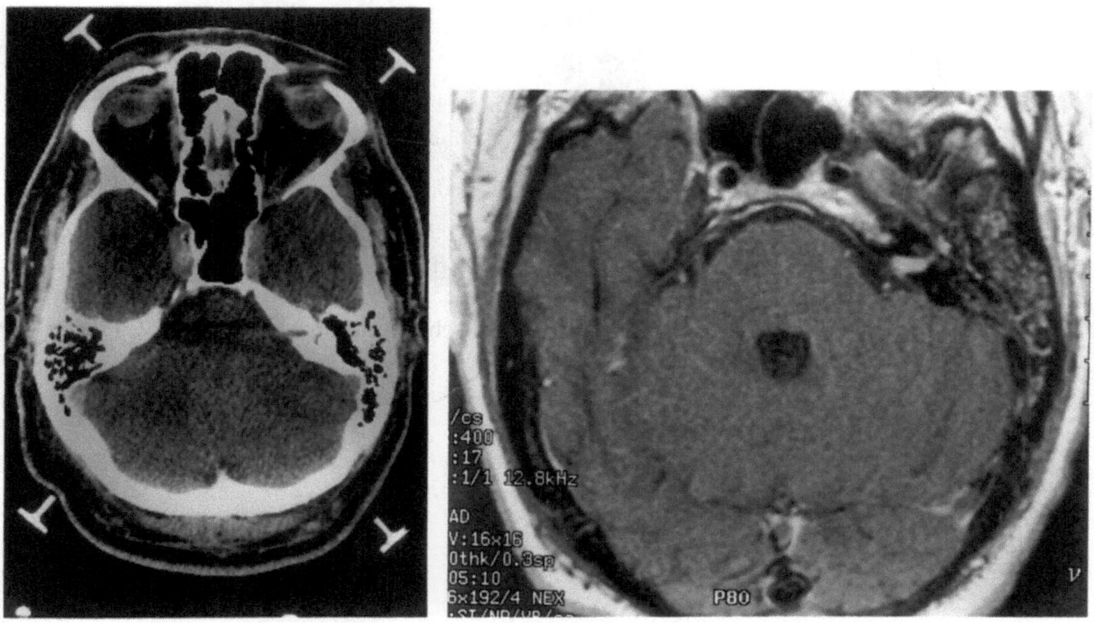 B

Figure 4.15. Intracanalicular acoustic schwannoma shown on CT scan (A) versus MRI (B).

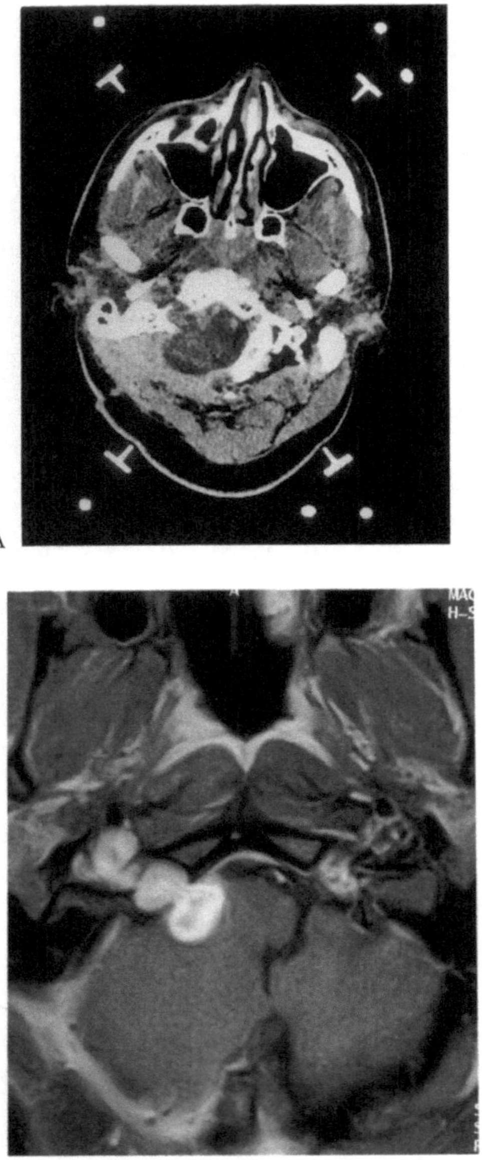

A

B

Figure 4.16. Jugular foramen schwannoma shown with CT scan (A) versus MRI (B).

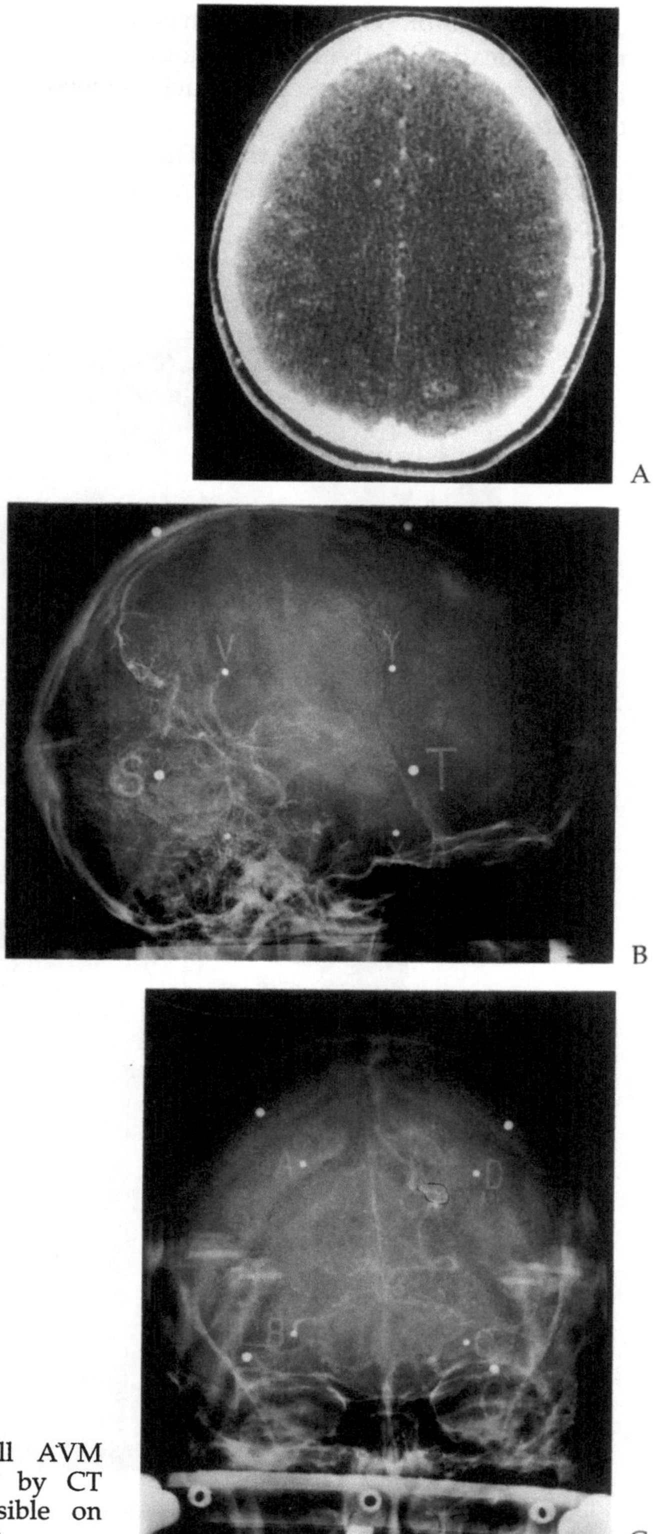

Figure 4.17. Small AVM visualized poorly by CT scan (A) but visible on angiography (B–C).

When in doubt, obtain a CT with contrast prior to radiosurgery ring application. If the lesion is not adequately imaged, plan to include another imaging study, such as stereotactic angiography or stereotactic MRI.

4. *Lesion better seen on CT scan.* The imaging characteristics of a lesion occasionally, favor CT scan. Figures 4.18 and 4.19 show two examples.

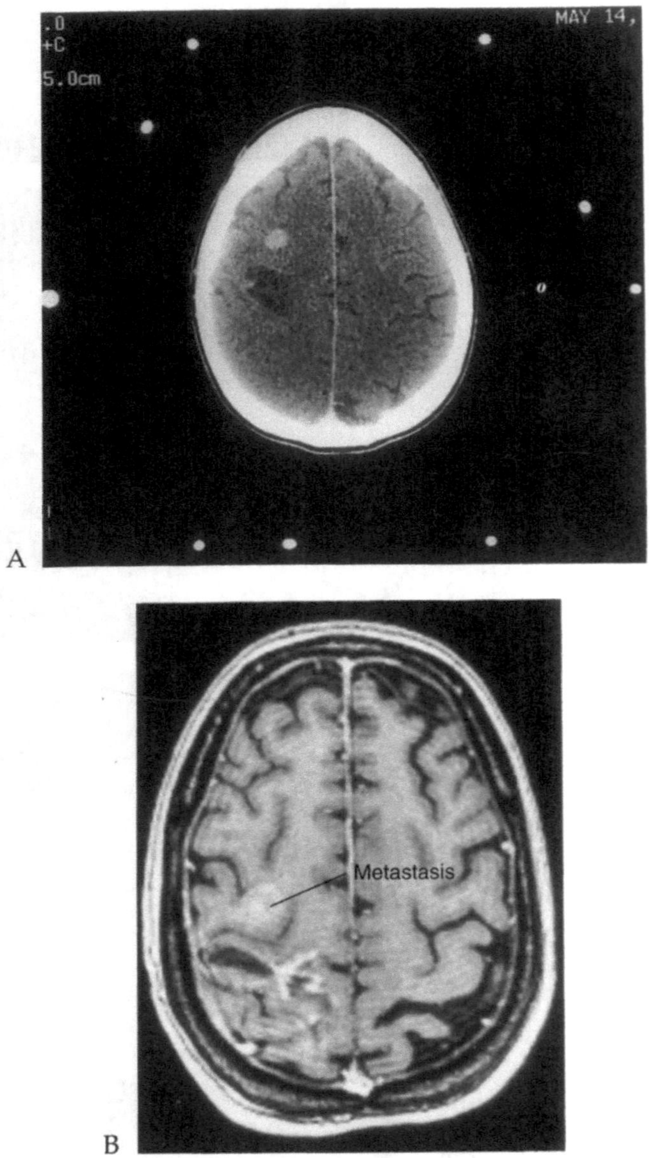

Figure 4.18. Metastasis from melanoma seen better on CT (A) than on MRI (B).

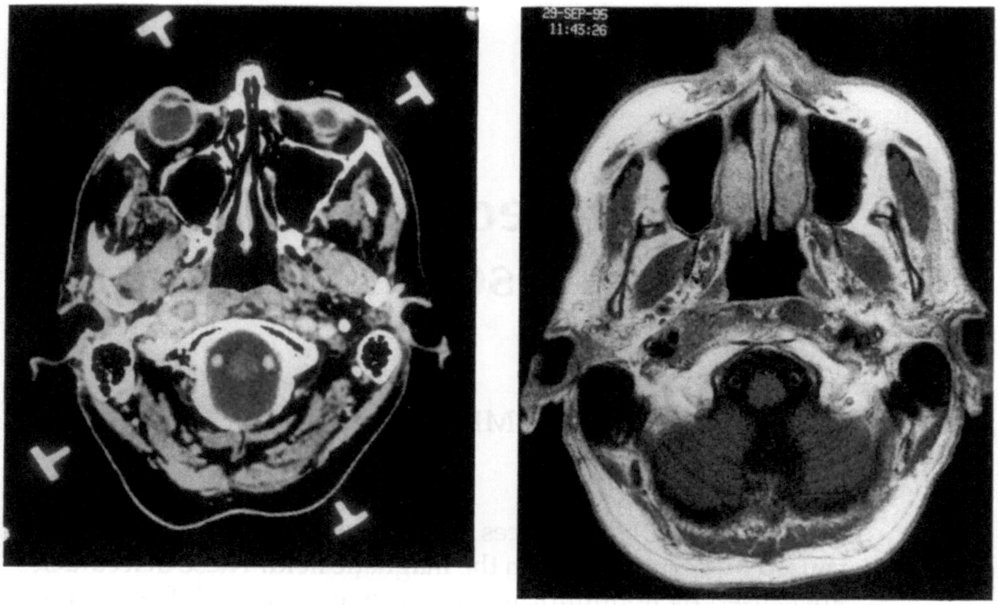

A B

Figure 4.19. Recurrent retropharyngeal squamous carcinoma seen better on CT (A) than on MRI (B).

5. *Thoracic kyphosis.* The presence of severe thoracic kyphosis occasionally, results in difficulty attaching the stereotactic head ring to the CT bracket. This difficulty can be minimized by relatively high ring placement, if possible, as well as by elevation of the patient's hips and lower back (Fig. 4.20).

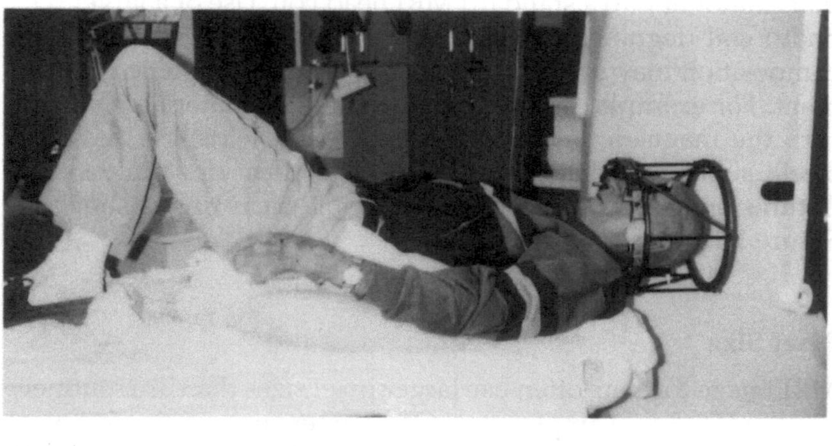

Figure 4.20. Severely kyphotic patient being CT scanned with hips raised to allow CT bracket attachment.

5

Stereotactic Magnetic Resonance Imaging

Pitfalls of Stereotactic MRI

Image Accuracy

MRI, unlike CT, produces images with spatial inaccuracies caused by perturbations in the magnetic field. These inaccuracies are caused by nonuniformity of the magnetic field. Perturbation of the magnetic field uniformity by the stereotactic frame and localizer may occur, as can perturbations of the magnetic field unique to the individual patient. Perturbations unique to the individual are called *susceptibility artifacts*. All of these problems must be eliminated or corrected before a scan can be incorporated into a stereotactic procedure.

Head Coils

One problem in acquiring stereotactic MRI is the size of the stereotactic frame and its compatibility with the MRI head coil. Certain frames, such as the BRW frame, have large diameters that cannot fit into a standard MRI head coil. Use of a larger (i.e., body) coil degrades image quality. In addition, the head ring composition may not be compatible with the magnetic environment. For example, if the head ring contains materials that perturb the magnetic or radiofrequency (RF) fields (i.e., stainless steel), significant distortions occur. The effect of stainless steel fixtures on an otherwise MRI-compatible head ring is shown in Figure 5.1.

Pixel Size

MRI image formats often use larger pixel sizes than are routinely used in CT. For example, most CT scanners use a 512×512 pixel image format. The scan diameter is 30 to 34 cm to include all CT localizer points, which results in a pixel size of 0.59 to 0.66 mm.

(A)

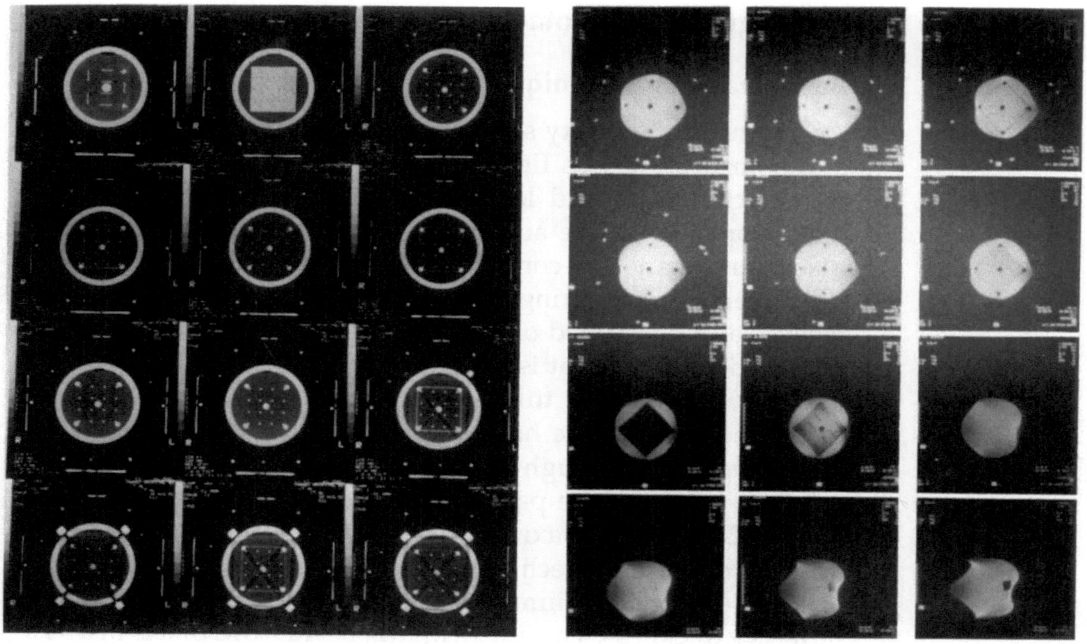

(B)

Figure 5.1. (A) An undistorted CT image of an MRI-compatible head ring with stainless steel fixtures. Note the artifact in the plane of the steel fixtures in the lower slices shown. (B) The effect of these stainless steel fixtures in an MRI image. Note the gross distortion and warping that occurs.

MRI most often uses a 256 × 256 pixel format. Thus, for the same image diameter (i.e., 30 to 34 cm), pixel size for MRI is doubled to 1.18 to 1.33 mm. State-of-the-art radiosurgery offers localization, treatment planning, and treatment delivery with accuracies less than 1 mm. Imaging targets with pixel sizes larger than 1 mm seriously degrades the overall accuracy of the radiosurgery procedure. Using this larger image format, a 1-pixel shift, caused by one or more of the potential image perturbations previously mentioned, causes a targeting error in excess of 2 mm. Because the gradient between therapeutic and nontherapeutic dose delivery (i.e., the typical distance in which the dose decreases from 90% to 50% of the target dose) for stereotactic radiosurgery is on the order of 2 mm, this is an unacceptable targeting error.

Lesions Poorly Seen on MRI

Lesions are rarely seen more clearly on CT than they are on MR images. This has occurred with skull base lesions and with certain hemorrhagic metastases (see Figs. 4.18 and 4.19).

Paradigm for Stereotactic MRI Acquisition

Basic Scanning Technique

Achieving high-quality stereotactic MRI for treatment planning at our institution was first addressed by the use of a modified stereotactic frame and localizer system that minimizes image distortion. The image acquisition with this technique requires placement of an MRI-compatible, modified BRW head ring. After placement of the ring, the patient is transported to the MRI facility, and positioned on the imaging couch with the head in a modified localizer that is coupled to a specially tuned MRI coil (Fig. 5.2). The coil is tuned such that the image is minimally perturbed by both the head ring and the localizer system. The coil is also small enough to maintain image quality.

Before imaging, the patient is injected with gadolinium contrast (0.2cc/kg) to maximize lesion enhancement on MRI images. The scanning technique employs T_1-weighted (T_R = 500 and T_E = 20) gadolinium-enhanced image acquisition using 3-mm slice thickness. The image matrix is 512 × 512 pixels in a 34.5-cm field of view so that the pixel size (AP and lateral) is identical to that used in CT (0.6 mm). Slice thickness is limited by the time of acquisition for the smaller pixel sizes used, although it is possible to use 1.5-mm slice thickness in a limited study area.

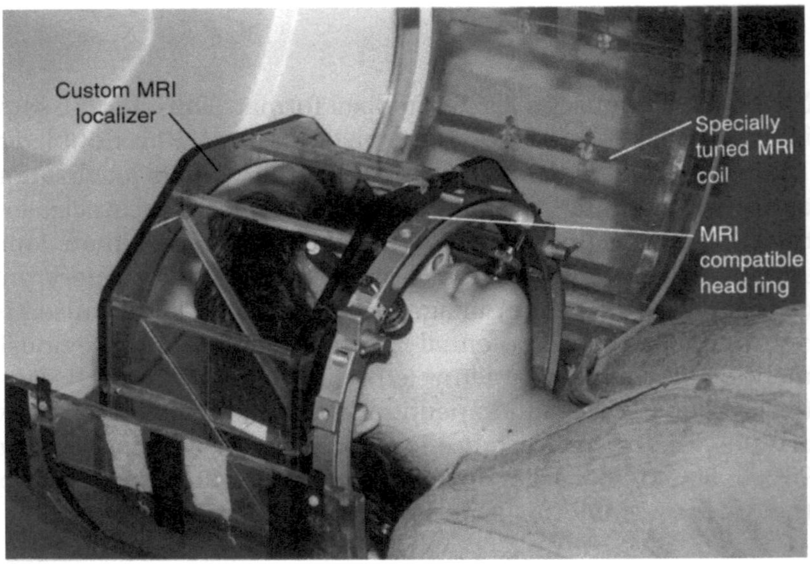

Figure 5.2. MRI-compatible stereotactic localization system includes an MRI-compatible head ring, a custom or modified MRI-compatible localizer, and a specially tuned MRI coil that allows the larger field of view for the localizer with minimal image degradation.

Similar to the procedure before CT image acquisition, a scout image (i.e., central sagittal MRI slice) is used to identify the area of interest for the MRI images. Treatment planning allows interchange of the dose plan between CT and MRI imaging so that the treatment plan based on MRI is easily compared with or overlayed onto the CT scan, and vice versa.

Image Fusion

A second way to obtain stereotactic MRI images for the treatment planning database is through the use of computer-generated image-annealing software programs, commonly termed *image fusion*. This technique allows MRI images acquired without the stereotactic head ring to be used for treatment planning. The MRI scan used for image fusion is routinely obtained the day before treatment. Images acquired for image fusion use the standard diagnostic MRI head coil. The scan is not limited to the area of interest, but includes the entire head. The scan technique uses volumetric image acquisition with a modified T_1-weighted sequence (SPGR, $T_R = 19.2$ and $T_E = 4.2$). This technique allows rapid image acquisition so that movement during the MRI is minimized. Image fusion eliminates many of the hardware incompatibility problems involved with using MRI for treatment planning. The volumetric scan technique also allows 1-mm slices, identical to the CT technique. Image quality is identical to that used for diagnostic MRI scanning. Figure 5.3 illustrates an example of an image fusion treatment database being compared with a CT treatment database before beginning treatment planning.

Developments in software annealing programs have allowed more facile use of MRI in the treatment planning process. Image distortion is reduced by removing hardware requirements of direct stereotactic MRI scanning. CT images are needed to perform the annealing process. Review of the MRI and its agreement or match with the CT database is essential. Treatment planning using MRI is likely to increase in the coming years because many lesions are poorly or suboptimally imaged on CT scans (see Figs. 4.15 and 4.16).

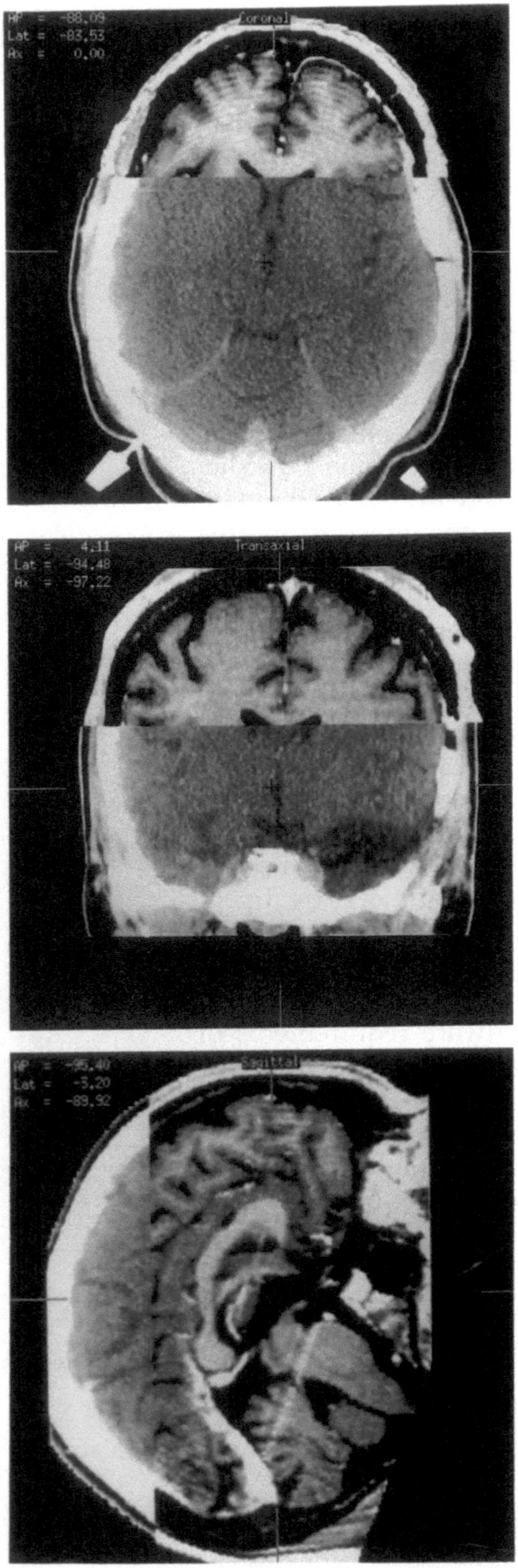

Figure 5.3. Image fusion software showing comparison of fused CT and MRI scans in th axial (A), coronal (B), and sagittal (C) planes. Note the MRI aligns well to the gyri, ventricle outline, and bony anatomy at the junction line in each image. This line may be scrolled up and down to confirm the quality of the image fusion, Attention to appropriate image annealing is critical to assure that the nonstereotactically acquired MRI is registered accurately to the stereotactic CT scan.

6

Radiosurgery Treatment Planning

Goals of Radiosurgery Treatment Planning

An ideal radiation treatment plan would deliver 100% of the desired dose to the treatment target and none to the normal brain. This is not possible in reality, but the primary goal of radiosurgery treatment planning is to achieve a plan that conforms to the target as closely as possible, as defined by radiation isodose shells (explained in detail shortly) within the plan. A number of treatment planning tools are available for adjusting the shape of treatment isodose shells so that they fit even highly irregular target shapes. Regardless of its shape, the entire target must be treated within the prescription isodose shell, with as little normal brain included as possible.

Another goal of dose planning is to adjust the dose gradient such that critical brain structures near the target receive the lowest possible dose of radiation. In addition, most linac radiosurgeons strive to produce a treatment dose distribution that maximizes uniformity (i.e. homogeneity) of the dose throughout the entire target volume.

This chapter begins with a discussion of basic radiation physics—just enough so that the reader can understand the theoretical basis of treatment planning. The major treatment planning tools will then be introduced. The University of Florida algorithm for using these tools will be discussed. Finally, an assortment of treatment plans will be presented in detail.

Basic Physics for Linac Radiosurgery

Linacs were developed simultaneously in the United States and Great Britain in the 1950s. These devices accelerate electrons to nearly the speed of light. The accelerated electron beam is aimed at a heavy metal alloy target, and the resulting interactions between the electrons and target produce X-rays, which are a form of photon energy. These photons, like light, can be collimated

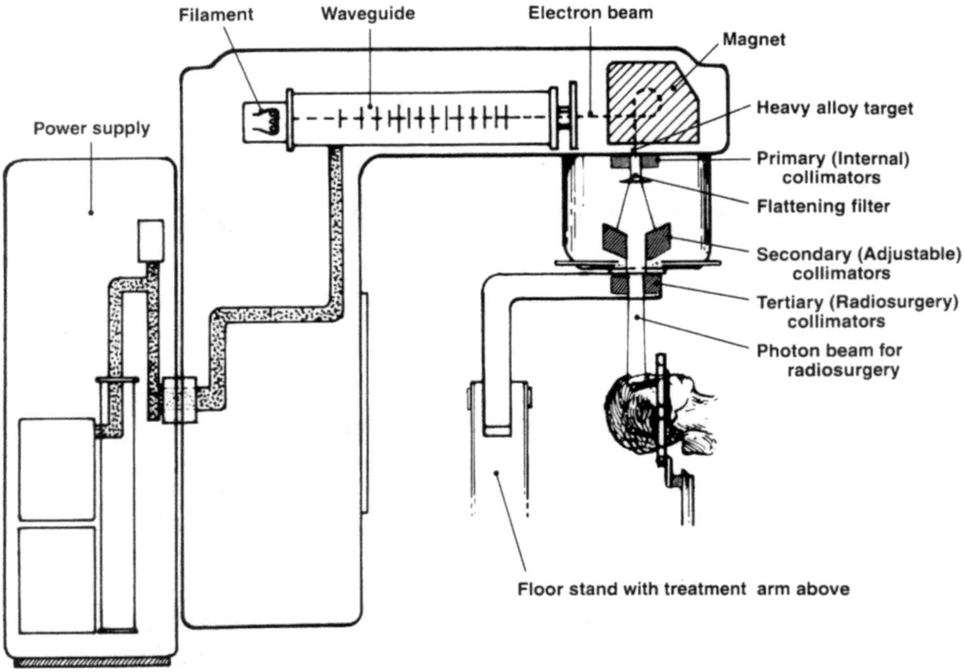

Figure 6.1. The linear accelerator is a complex machine capable of producing X-rays for radiation treatments. A large amount of energy is generated by the power supply, which then powers the filament shown. This causes electrons to be emitted by the filament, which are in turn accelerated to higher energies using a (microwave) wave guide. The electrons are then changed in direction by the magnet so that they impact on a heavy metal alloy target. This results in X-ray production that can then be collimated or shaped by both primary and secondary collimators (i.e., jaws) within the linear accelerator head. This beam is further collimated for radiosurgery by the tertiary radiosurgery collimator. In the University of Florida system this tertiary collimator is within the treatment arm of the floor stand for reasons that will be discussed in Chapter 8.

and focused on a patient. Over the ensuing decades, linacs have become the favored treatment devices for conventional radiation therapy because of their durability and versatility (Fig. 6.1).

As a photon beam traverses an absorbing material, its intensity decreases exponentially. Figure 6.1 shows a radiation beam entering a patient. As the beam progresses in depth, it loses intensity. To display the dose delivered at any point within the brain, lines delineating the edge of various dose percentages can be displayed on a two-dimensional image. The points at which 90%, 80%, 70%, 60%, 50%, and 40% of the maximum radiation dose will be found are shown in Figure 6.2A and B. Because the lines connect all points of the same dose, they are known as *isodose*, or same dose, lines. Although they appear as lines on the two-dimensional display, they represent in reality three-dimensional shells surrounding a volume treated to a dose (Fig. 6.2C). All

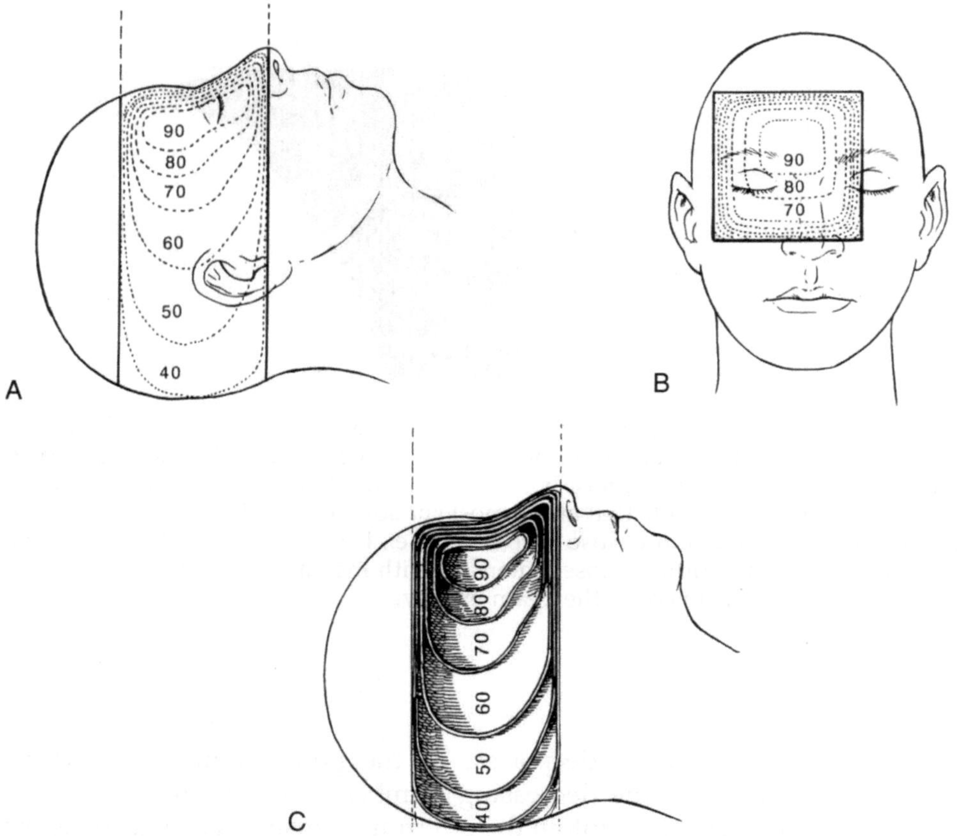

Figure 6.2. The course of a radiation beam through a model patient. (A) In the lateral view the dose intensity falls with depth. The 90%, 80%, 70%, 60%, 50%, and 40% isodose lines are shown in this two-dimensional representation of the center of the beam in a sagittal plane through the patient. (B) Schematic representation of the same patient's radiation isodose lines as they would look in a coronal plane at the level (depth) of the lateral canthus (as shown in A). (C) Three-dimensional rendering of isodose shells as they would look on the model patient in the sagittal plane.

subsequent dose displays in this chapter show radiation isodose lines as a percentage of the maximum dose.

The absorption of a photon beam traveling through tissue can be broken down into two regions (Fig. 6.3). The first, called the *buildup region,* starts as the photon beam enters the patient. Over a relatively short distance, the dose deposited increases until the maximum dose is reached. This buildup region exists because the photon beam does not deposit energy directly into tissue. Instead, it interacts with the tissues to create high-energy electrons and scattered photons of lower energy. The electrons actually deposit most of the energy into the tissue.

As the photon beam creates high-energy electrons, photons are removed from the beam. Thus, as electron intensity increases,

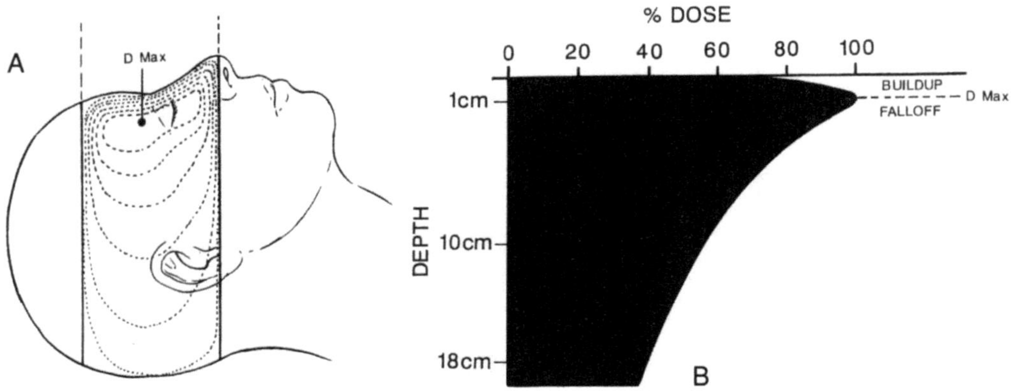

Figure 6.3. A photon beam may be divided into a buildup region and a falloff region. (A) A magnified isodose plot that illustrates both regions on a patient as they would look in the sagittal plane. This is correlated graphically in (B). (B) A depth–dose curve that illustrates the relationship of dose to depth. Note that at the most shallow depths the dose is lower (i.e., buildup region), and then a maximum dose is reached (i.e., Dmax). This is followed by the falloff region where the beam gradually loses intensity with increasing depth. The shape of each depth–dose curve is determined by the beam energy.

photon intensity decreases. At the point of maximum dose, called *Dmax,* the decreasing number of remaining photons is insufficient to maintain the electron intensity, which also begins to decrease. At the skin surface, the radiation dose delivered by a single fixed beam from a linac is usually between 15 and 30% of the maximum dose for the beam, and the Dmax is reached at a depth specific to the beam's photon energy. The decrease in dose at greater depths than Dmax marks the second portion of the absorption curve, known as the *falloff region*.

Because the depth–dose curve is relatively fixed for a given photon-beam energy, a single fixed beam would never deliver a maximum dose to the target tissues unless the target happened to lie at the depth of Dmax. As this is rarely the case, a method of concentrating dose at depth must be used. Use of multiple beams, each with a unique entrance and exit pathway, yet all directed at a single target, is one method of achieving this. This is the basis of modern radiotherapy treatment planning, which generally uses two to four fixed, coplanar radiation fields defined by collimators within the linac and further shaped by custom blocking. The concept of using multiple beams is extended by the radiosurgery treatment paradigm used for linac and gamma knife systems. Gamma knives use 201 separate cobalt sources, all aimed at one target. Linacs use multiple, noncoplanar arcs of radiation, all focused on one target (Fig. 6.4). In

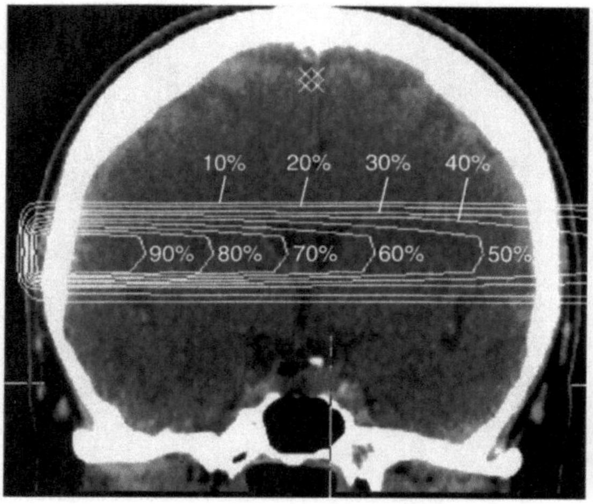

(A)

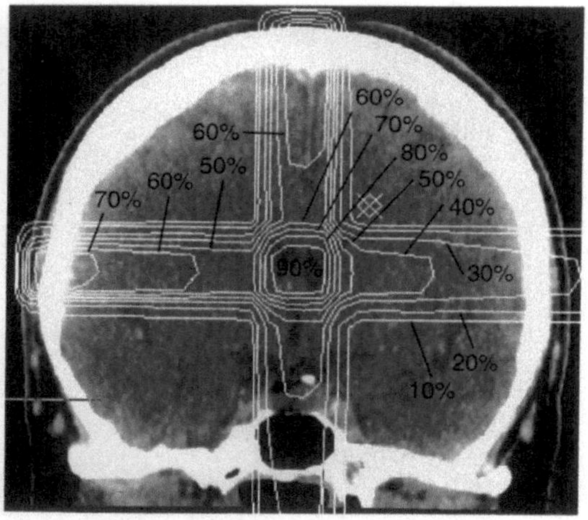

(B)

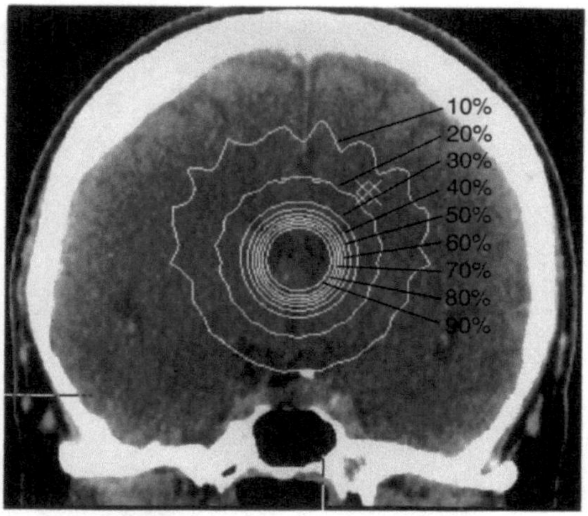

(C)

Figure 6.4. Illustration of the change in dose concentrations by the addition of multiple radiation beams. Doses are represented as isodose lines (labeled) in a coronal plane for a single field (A), three fields (B), and radiosurgical treatment (nine arcs) (C). The radiosurgery paradigm is the equivalent of hundreds of radiation beams. Note the differences in areas encompassed by the labeled isodose lines based on number of fields.

the stereotactic paradigm, the equivalent of hundreds of radiation beams are focused on a selected target.

For example, take a single beam entering a skull, with the intensity of the beam reflected as isodose lines (Fig. 6.5). For this example, assume that the skull cross-section has a 24-cm diameter with the target in the center (i.e., 12-cm deep from all surfaces). The Dmax of a single 6 MV photon beam is approximately 1.5 cm from the surface, and the dose at the target will be approximately 55% of the maximum dose (Fig. 6.5A). If a second beam entering the skull at right angles to the first is added, then the dose at the center of the skull will be the summation of the two beams: two times 55%, or 110% (Fig. 6.5B). If six more beams are added, for a total of eight beams, as shown in Figure 6.5C, then the dose at the target will be eight times 55%, or 440% of the maximum dose from any single beam; and if the number of

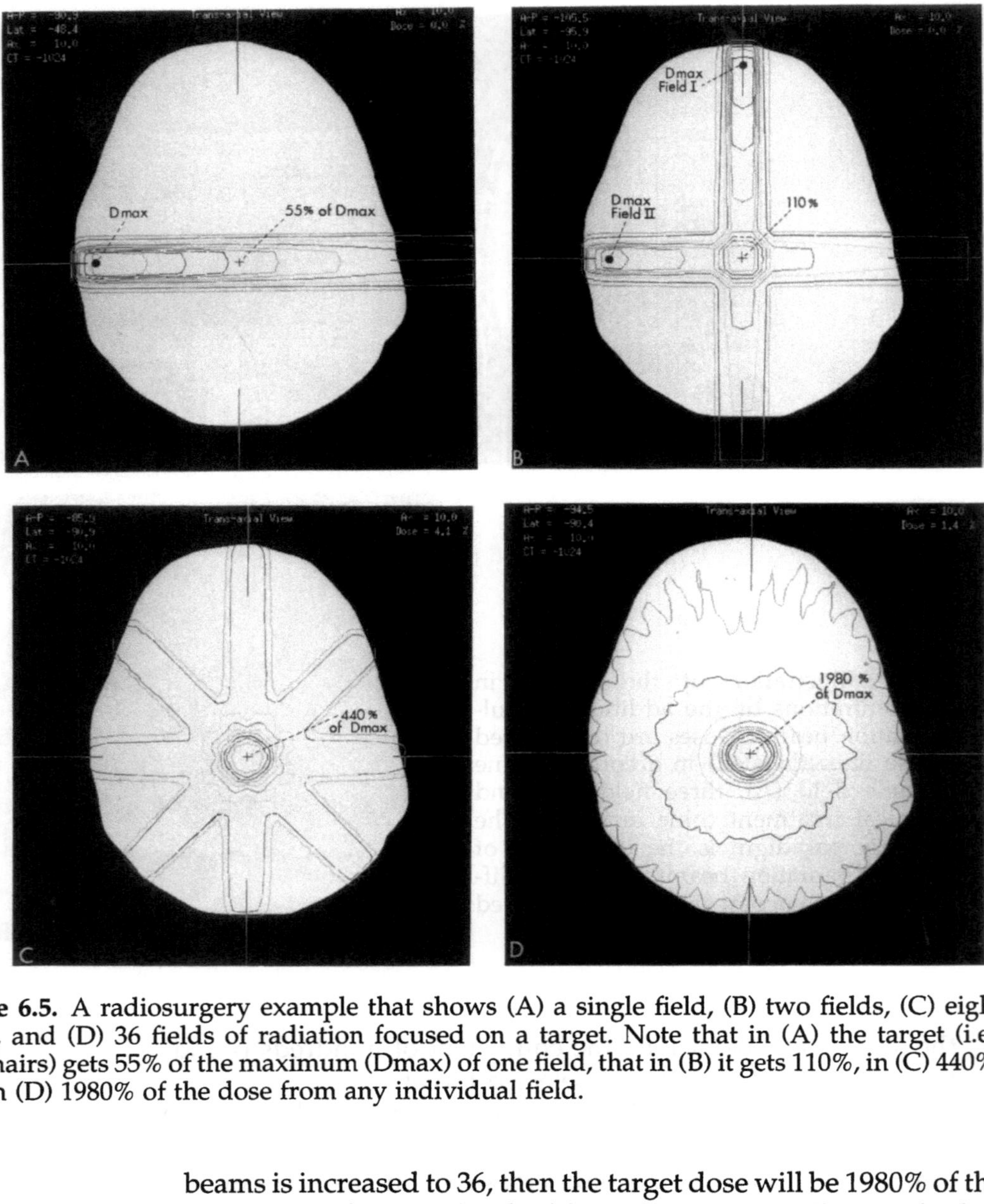

Figure 6.5. A radiosurgery example that shows (A) a single field, (B) two fields, (C) eight fields, and (D) 36 fields of radiation focused on a target. Note that in (A) the target (i.e., crosshairs) gets 55% of the maximum (Dmax) of one field, that in (B) it gets 110%, in (C) 440%, and in (D) 1980% of the dose from any individual field.

beams is increased to 36, then the target dose will be 1980% of the maximum of any individual beam (Fig. 6.5D). This method of summing the dose contributions from many beams or arcs of radiation at one target is the basis of radiosurgery.

In practice, a set of beam attenuation curves is determined for each size of collimator (beam-shaping device) used in radiosurgery. In this way, the dose contributed by each radiation beam to a target at a given depth is defined. Beam diameters of 5 to 40 mm are typically available for standard linac radiosurgery. The distance each beam will travel through tissue before it reaches the target is readily computed during treatment planning from the reconstructed CT scans (see Chapter 4). Using the predetermined, collimator-specific attenuation data and the known depths to target, dosimetry software programs can rapidly compute and display the isodose information for any proposed combination of radiation beams and target dose desired.

Treatment Planning Tools

The following treatment planning tools are commonly used in radiosurgery treatment planning: arc weighting, altering arc stop and start angles, and the use of multiple isocenters.

Arc Weighting

Arc Elimination

In general, we begin treatment planning by directing nine equally spaced arcs of radiation at the center of the target (Fig. 6.6). Each arc span is 100 degrees, and each arc is spaced 20 degrees from its neighboring arcs. This results in a spherical dose distribution, with the dose falling off equally in all directions.

Many radiosurgical targets are not perfectly spherical; rather, they are shaped more like an elongated sphere (ellipsoidal). It is relatively easy to change the spherical dose distribution into an ellipsoidal distribution with linac radiosurgery systems. All that is required is to eliminate the arcs (i.e., reduce their weight to zero) that are most perpendicular to the long axis of the ellipsoid. For example, in Figure 6.7A, the target is elongated in the superior–inferior direction. By eliminating the four arcs of radiation that are most horizontal in orientation, the dose distribution is elongated in the required direction, which is the principal direction of the remaining arcs. As another example, consider Figure 6.7B. Here, the target is elongated in a lateral direction. This shape can be matched by eliminating the three most vertical arcs of radiation.

A corollary to this method is that elimination of an arc reduces the spread of radiation in the principal direction of the eliminated arc. For example, when treating an acoustic schwannoma, it is desirable to minimize the radiation given to the medially located brain stem (see Case Example 2). This is effectively accomplished by eliminating the four most horizontal arcs, which would enter or exit through the brain stem. The radiation dose is elongated into the superior–inferior direction, where less critical structures are located, and the spread of radiation over the medially located brain stem is reduced. As an alternative, when treating a pituitary tumor, the adjacent critical structure is the optic chiasm, which is located superior to the tumor. Eliminating the

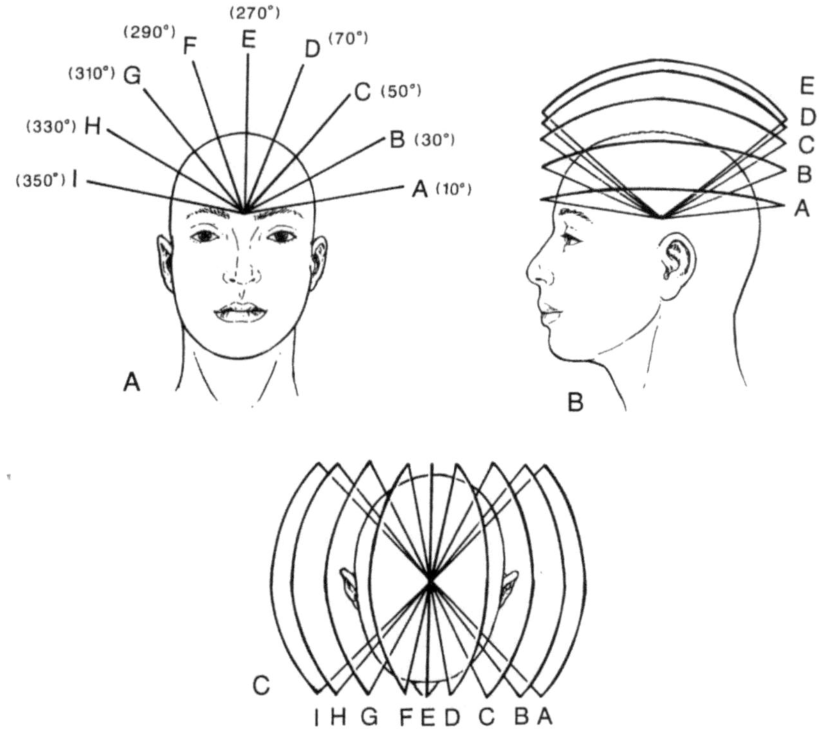

Figure 6.6. Radiosurgery treatment planning at the University of Florida begins with nine equally spaced arcs of radiation. (A) Coronal orientation of the nine arcs. Treatment table angles (in degrees) to deliver each arc are noted in parentheses, and each arc is lettered A–I. (B) Lateral view showing the arc orientations. (C) Superior view of the nine arcs lettered to coincide with each table angle shown in (A).

three most vertical arcs reduces the spread of radiation in the superior–inferior direction, but increases the spread in the lateral direction where less sensitive structures (e.g., cavernous sinuses) are located. This dose gradient modification might be called the "Jell-O® principle" of radiosurgery. If one compresses a spherical Jell-O® mold in one direction, it will elongate in another. By using the tool of arc elimination and by understanding the Jell-O® principle, both the primary goal of conforming to the target and the secondary goal of sparing critical structures may be approached. In radiosurgery treatment planning designed to avoid critical structures, the elongated portion of the Jell-O® mold is in the direction of the least critical structures.

The bottom line. For any target that is approximately ellipsoidal and has its principal axis anywhere in the coronal plane, treatment can be planned by eliminating the arcs that are most per-

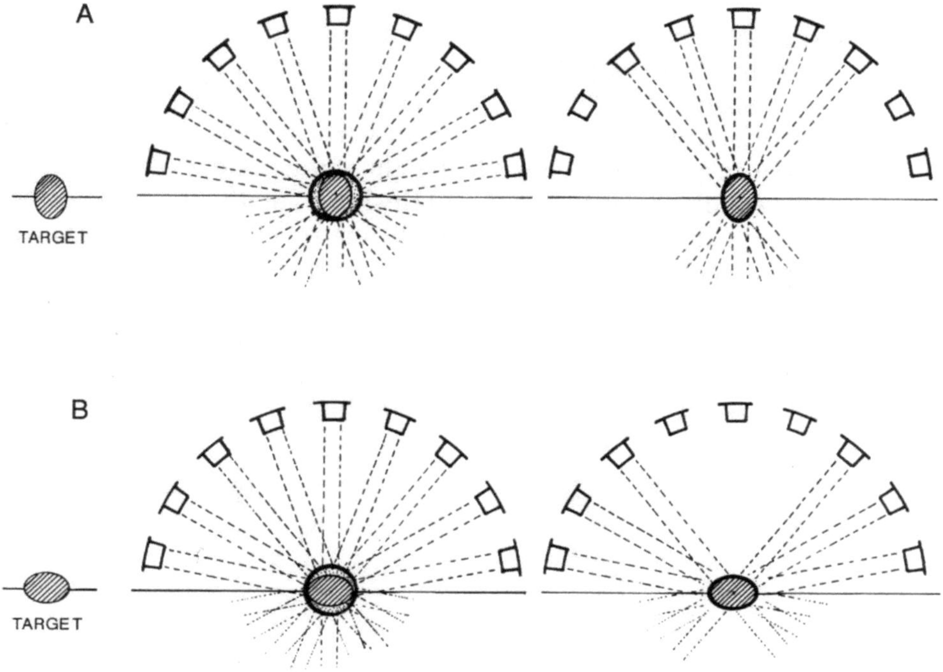

Figure 6.7. (A) An ellipsoidal target in the coronal plane with a principal axis in superior–inferior direction is represented by the cross-hatched region. The spherical dose distribution created by the nine-arc radiosurgery starting point is shown with a bold line. Each arc is represented by a schematic collimator, and each entering radiation beam by dotted lines. Note that the isodose line does not conform to the target (see stippled region surrounding the target but within the bold line). By eliminating arcs of radiation perpendicular to the long axis (i.e., the more horizontal arcs) of the ellipsoidal target, the dose distribution (bold line) conforms to the target, and the stippled area is eliminated. (B) An ellipsoidal target in the coronal plane with a principal axis in the medial–lateral direction. The target is again cross-hatched and the dose distribution of the minimum-size nine-arc treatment plan is adjacent. The stippled area illustrates the lack of conformality of the isodose line (bold) to the target. By eliminating arcs of radiation perpendicular to the principal axis of the target (i.e., the more vertical arcs), the conformality is significantly better and the stippled area outside the target is eliminated.

pendicular to its principal axis. In addition, arc elimination maximizes the dose gradient in the direction of the eliminated arc. This strategy is the most frequently used treatment planning method in linac radiosurgery and will, by itself, result in a conformal treatment plan for many radiosurgical lesions.

Differential Collimator Sizes

The overall weight of an arc can be changed (i.e., increased or reduced), rather than completely eliminated, by increasing or reducing the size of the beam (i.e., collimator size) used for the arc. Referring back to Figure 6.7A, notice that the diameter of the target is greater in the vertical direction than it is in the horizontal direction. The horizontal beams of radiation therefore encounter a taller target than do the vertical beams of radiation. In other words, the "beam's eye view" perceives a bigger target from horizontal than it does from vertical (Fig. 6.8). To look at this from a different view, the height of the isodose configuration is most influenced by the size of the horizontally oriented beams, whereas the width (i.e., lateral diameter in the axial plane) is most determined by the size of the vertically oriented beams.

Considering radiation beams as overlapping tubes is another way to think about this. The dose distribution is shaped by the intersection of all of these tubes. By placing smaller tubes through the vertical direction rather than through the horizontal,

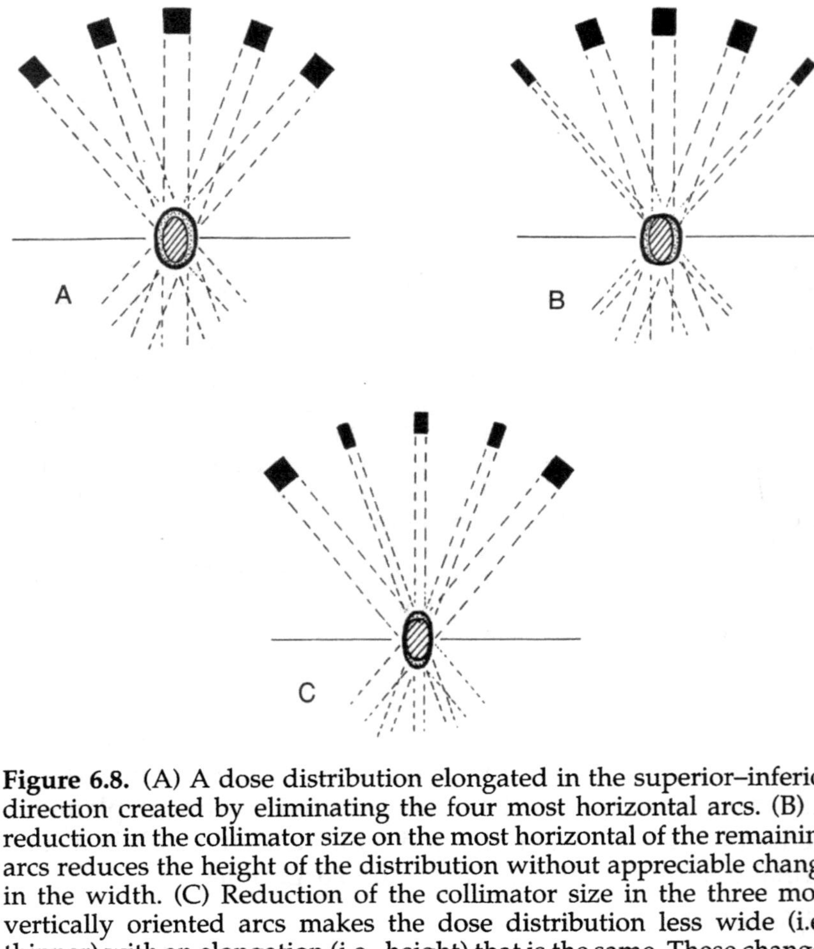

Figure 6.8. (A) A dose distribution elongated in the superior–inferior direction created by eliminating the four most horizontal arcs. (B) A reduction in the collimator size on the most horizontal of the remaining arcs reduces the height of the distribution without appreciable change in the width. (C) Reduction of the collimator size in the three most vertically oriented arcs makes the dose distribution less wide (i.e., thinner) with an elongation (i.e., height) that is the same. These changes in differential collimator sizes allow for fine tuning of the ellipsoidal distributions to account for different ratios of the principal to nonprincipal axis. Changes in collimator sizes have relatively little effect on the lower isodose lines in a radiosurgery dose distribution compared with arc elimination.

the overlap will be less in the lateral direction, but it will maintain the spread of distribution in the superior–inferior direction. If this model is extended and one has a target as shown in Figure 6.7A with arc elimination complete, then a further change in the distribution will occur if the most horizontal of the remaining collimators are reduced in size (Fig. 6.8A,B). When this occurs, the distribution becomes less elongated in the superior–inferior direction because the overlapping of the beams is decreased in height because the height is most controlled by the horizontal beams. Decreasing the size of the more vertically-oriented arcs will, conversely, diminish the lateral spread of the overlapping tubes of radiation and will create a distribution of the same height that is slightly more narrow (Fig. 6.8C). Hence, as an alternative to arc elimination, different collimator sizes can be used on different arcs. This strategy results in slightly more or less severe elongation (i.e., ratio of principal to nonprincipal axis) of the treatment isodose configuration, with much less change in elongation of the lower isodose lines into surrounding tissue.

As a practical matter, differential collimator sizes and arc elimination are often used in combination to fine-tune the shape of the treatment isodose curve. This is especially useful when arc elimination is used primarily to reduce irradiation to surrounding structures (see Case Example 2).

The bottom line. Differential collimator sizes can be used to fine-tune ellipsoidal dose distributions. This will result in a more or less gently elongated treatment dose distribution, and much less elongation of lower isodose curves than what is seen with the arc elimination tool.

Altering Arc Start and Stop Angles

We typically use arcs spanning 100 degrees, starting 30 degrees off superior vertical and ending 50 degrees off inferior vertical (Fig. 6.9A). This results in the particular AP orientation (sagittal plane) of the isodose curves shown in Figure 6.9B. The target lesion sometimes has a significantly different AP orientation. For example, in Figure 6.10A, the lesion is tilted posteriorly compared with the standard isodose orientation. This lesion orientation can be more closely approximated by changing the superior vertical arc angles, thereby decreasing the total arc span and orienting the arc more closely to the target (Fig. 6.10B). The physical restrictions on start and stop angles prevent the practical application of this particular tool, unless the lesion has a primarily superior–inferior elongation, with relatively modest tilts in the AP direction.

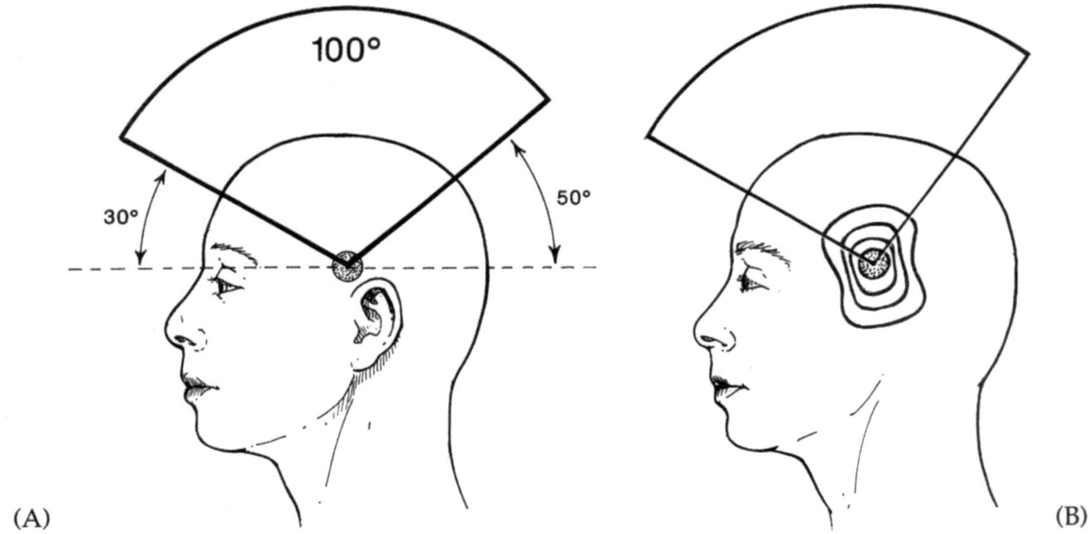

Figure 6.9. (A) Schematic representation of one of the standard nine arcs highlighting its orientation 30 degrees off superior–vertical, 50 degrees off inferior–vertical, and spanning 100 degrees. (B) The typical AP orientation of the dose distribution created by a standard nine-arc plan.

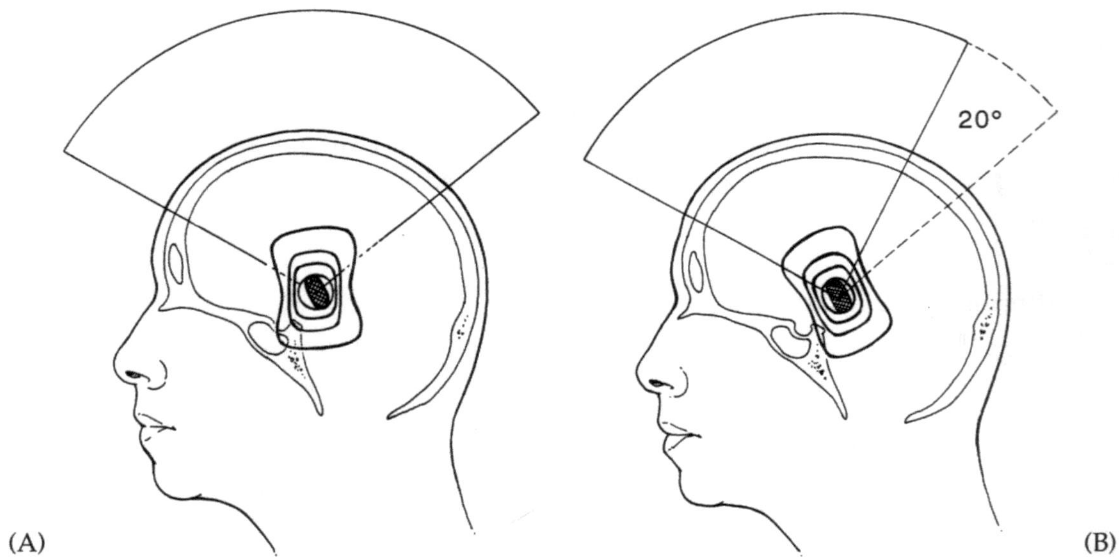

Figure 6.10. (A) The cross-hatched target has an AP orientation in the sagittal plane that is not spherical. The principal axis of this ellipsoidal shape extends from anterior–superior to posterior–inferior. It is elongated in the superior–inferior dimension. In (A) the standard nine-arc distribution does not conform to the shape of the lesion. To increase the dose conformality, an approximation of the arc span down the principal axis of the ellipse results in the improved distribution shown in (B). This is practically accomplished by deleting the most posterior 20 degrees (dotted lines) from each arc. This schematic illustrates one use of altering arc start stop angles.

Another application of this tool emerges because critical structures are sometimes immediately anterior or posterior to the target lesion. In this case, one would like to maximally reduce the dose gradient in the AP direction. This is readily accomplished by decreasing the arc span (Fig. 6.11). Narrow arcs effectively elongate the lower isodose lines in the superior–inferior direction, while narrowing them in the AP direction (another application of the "Jell-O® principle").

The bottom line. Arc start and stop angles can be adjusted so that the principal AP vector of the arc more closely matches the AP orientation of the target. In addition, reducing the arc span can be used to maximize the AP dose gradient.

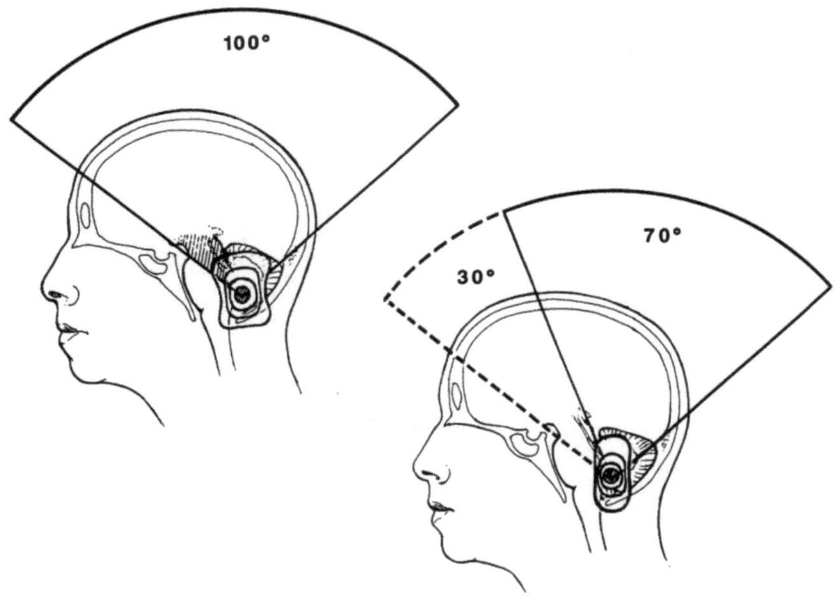

(A)

(B)

Figure 6.11. Altering the arc start stop angle can also improve the dose gradient. (A) Schematic representation of a target immediately posterior to the brain stem. This shows the typical dose distribution obtained with a standard nine-arc plan in the sagittal plane. *Note*: The anterior portion of this representative arc transgresses the critical neural structure, the brain stem. This area that we want to avoid is cross-hatched in the illustration. (B) By altering the anterior portion of each arc, the dose gradient is improved. In this case the anterior 30 degrees of each arc goes immediately through the brain stem and is removed. The resulting dose distribution (illustrated) has a superior dose gradient (sharper dose fall-off) anteriorly near the critical neurologic structure, the brain stem.

Multiple Isocenters

Arc weighting is used in treatment planning for lesions that are ellipsoidal in the coronal plane (Figs. 6.7 and 6.8). Changing the arc start and stop angles is used to match more closely the AP tilt of lesions that are primarily elongated in a superior–inferior direction (Fig. 6.10). If the lesion is ellipsoidal but primarily elongated in the AP direction (axial plane), however, then multiple isocenters must usually be used to produce a conformal plan (Fig. 6.12A). Likewise, if the lesion is nonspherical and nonellipsoidal, multiple isocenters must be used (Fig. 6.12B).

To use multiple isocenters, the three-dimensional shape of the lesion must be ascertained. This is done by viewing sequential axial CT or MRI images from the top to the bottom of the image,

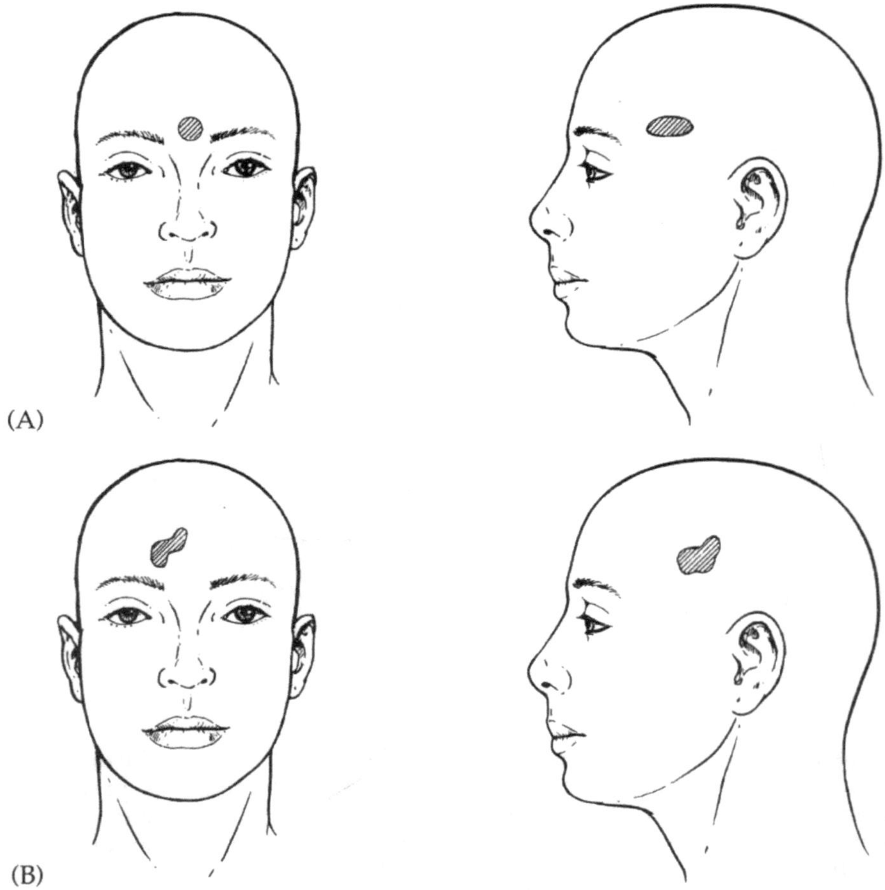

Figure 6.12. (A) A lesion that is significantly elongated in the axial plane, even though it is seemingly spherical in the coronal plane. This lesion would require multiple isocenters to achieve conformality with the AP elongation. (B) Highly irregular lesions in one or more planes of view (i.e., coronal and sagittal in this case) also require multiple isocenter treatment strategies.

or, optimally, a three-dimensional viewing window more easily accomplishes the same purpose. If the lesion is generally cylindrical, then two isocenters are used. If it is generally triangular, then three isocenters are used. If it is generally shaped like a rectangular solid, then four isocenters are used. More isocenters are occasionally necessary to conform to a lesion of very irregular shape.

Once the three-dimensional shape of the lesion and the number of isocenters needed are determined, the isocenters must be positioned. This is accomplished by computer reconstruction of the plane through the lesion that contains its principal axis (Fig. 6.13). The isocenters are roughly positioned, in the appropriate orientation (i.e., line, triangle, or rectangle) on this plane. The

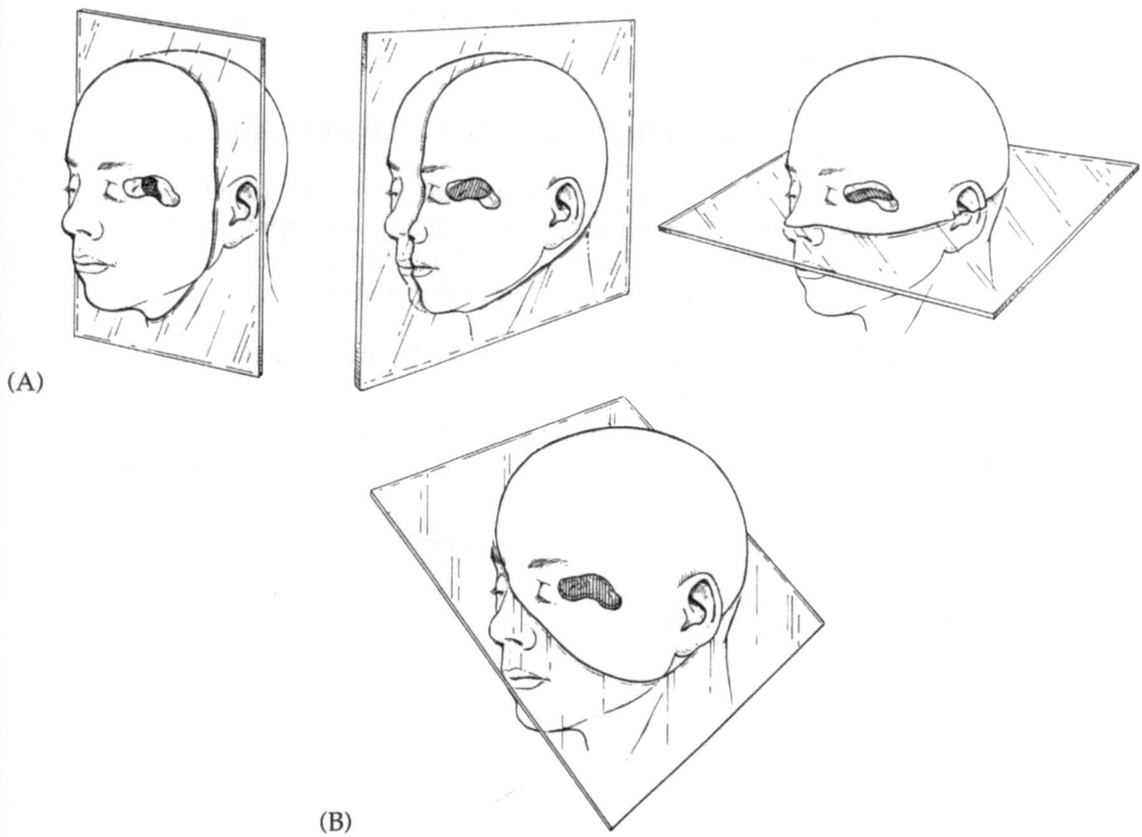

Figure 6.13. The lesion shown schematically in (A) from true coronal, sagittal, and axial perspectives, respectively, is growing along the temporal fossa. After reviewing the sequential axial slices it is clear that the principal axis or plane of the lesion is not in the true axial, coronal, or sagittal planes. By choosing three points within the lesion that most closely approximate the principal axis, a unique plane of view is obtained for the purpose of treatment planning. This plane is shown in relationship to the lesion in (B). The ability t· choose this three-dimensional viewing window aids in isocenter placement for complex three-dimensional targets.

approximate size of each isocenter is selected. A spacing chart, shown in Figure 6.14, lists most of our collimator sizes and the optimal spacing for combinations of multiple collimators. Using a computer spacing tool, the approximate isocenter positions are fine-tuned by moving them to the previously determined optimal spacing. We typically treat multiple isocenter lesions to the 70% isodose line, maintaining relative dose homogeneity.

With this strategy, multiple isocenter plans can be rapidly constructed. This interactive process is tremendously aided by fast computation times, as many adjustments are often necessary. Multiple isocenter planning requires training, practice, and real expertise to be applied optimally.

The bottom line. Multiple isocenters must be used for lesions that are either ellipsoidal in the AP direction or nonellipsoidal (i.e., irregular in shape).

University of Florida Treatment Planning Algorithm

The treatment planning tools have been described in detail. Next, a strategy is required for selecting and applying the tools to given lesion geometries. The algorithm generally used at the University of Florida is shown in Figure 6.15.

If the Lesion Is Adjacent to a Very Radiosensitive Neural Structure

The optic apparatus and the brain stem are of most concern because these are the most radiosensitive structures in the brain.

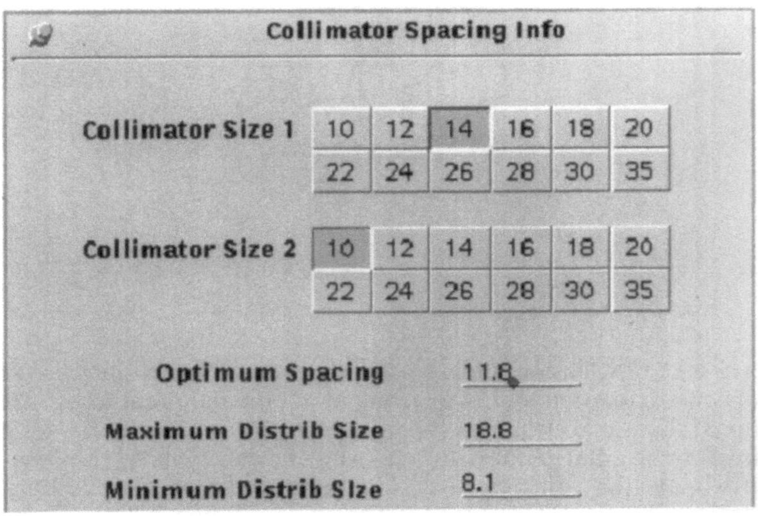

Figure 6.14. Spacing Chart. This chart allows the radiosurgeon to select collimator sizes for two adjacent isocenters. The computer then gives the optimal spacing of the isocenters so that they may be moved accordingly.

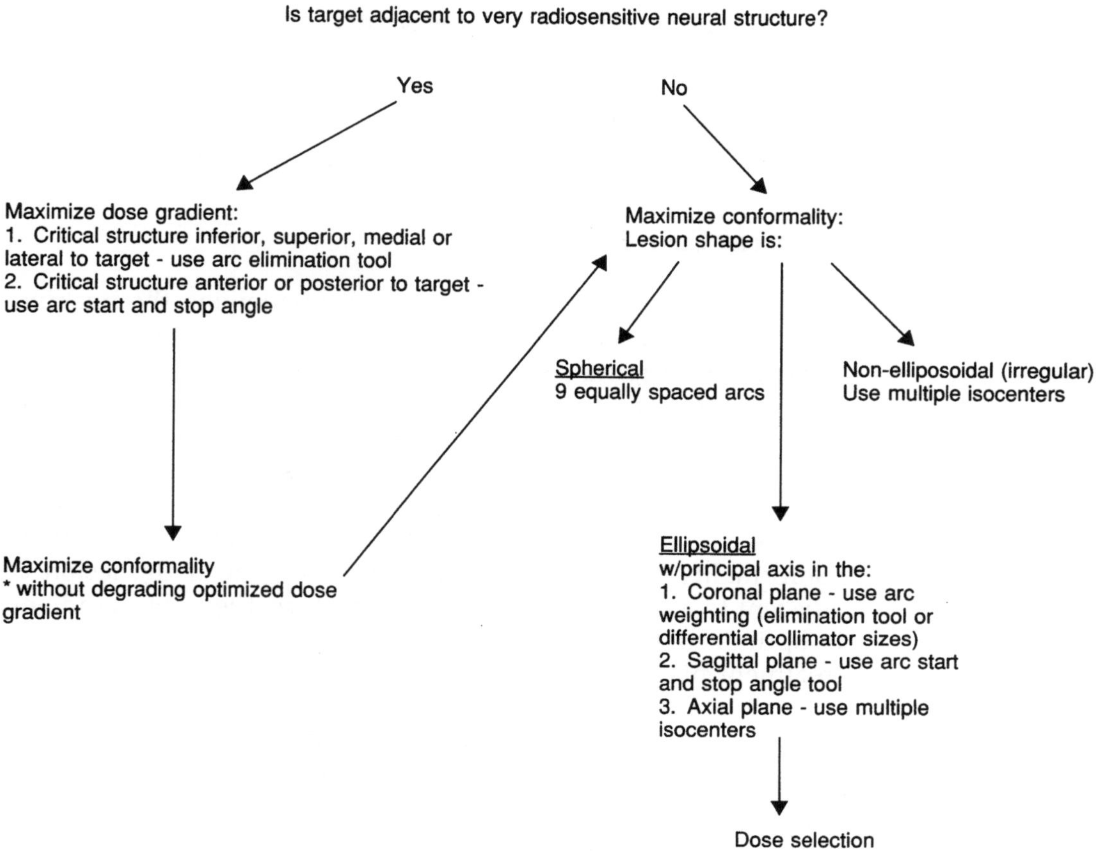

Figure 6.15. The University of Florida Stereotactic Radiosurgery Treatment Planning Algorithm.

If these structures are adjacent to the target lesion, then the primary strategy is to maximize the dose gradient in the direction of the critical structure. If the critical structure is inferior, superior, medial, or lateral, then the arc elimination tool is used to optimize dose gradient. If the critical structure is anterior or posterior, then the arc start and stop angle tool is used to optimize the gradient.

After Optimizing Dose Gradient for Sensitive Structures, or If the Lesion Is Not Adjacent to a Very Radiosensitive Neural Structure

Attention is given to producing a distribution that conforms to the lesion shape, but does not degrade the dose gradient achieved to protect sensitive structures.

If the lesion is spherical: Use nine equally spaced arcs.

If the lesion is ellipsoidal:

If the principal axis is in the coronal plane: Use the arc weighting tool.

If the principal axis is in the sagittal plane: Use the arc start and stop angle tool.
If the principal axis is in the axial plane: Use multiple isocenters.
If the lesion is nonellipsoidal: Use multiple isocenters.

Practical Case Examples

Before beginning any plan we begin by reviewing the imaging data thoroughly. We identify the center of the lesion in the axial, coronal, and sagittal planes, and trace the tumor contour into the computer for these planes. This process gives us an overall view of the lesion configuration and its location near or distant from critical neurologic tissue. This allows us to make the first decision along the algorithm path: Is the lesion adjacent to a radiosensitive neural structure? If the lesion is not, then conformality is immediately the primary goal; if the lesion *is* adjacent to a critical radiosensitive structure, then dose gradient is optimized first. For example, we will look at some actual treatment plans.

1. Spherical Arteriovenous Malformation

This patient initially had a seizure and was found to have an AVM in the region of the left trigone. The nine-arc treatment plan in Figure 6.16 shows the standard spherical distribution that we use to begin treatment planning. The dose distribution is spherical for the 80% isodose shell that is illustrated by the innermost isodose line in each of the axial, coronal, and sagittal planes. Subsequent, more peripheral isodose lines in each figure represent the 40%, 16%, and 8% isodose lines, respectively. Each arc is treated through 100 degrees of arc span. The orientation of each arc in relation to the coronal and axial planes is shown in Figure 6.16A. Although not common, this spherical distribution covers the lesion well with minimal normal tissue and represents an optimized plan.

2. Single Isocenter Acoustic Schwannoma

This patient initially had hearing loss and was discovered (with MRI scan) to have a left-sided acoustic schwannoma approximately 2.5 cm in diameter. We began our treatment planning with the standard nine-arc plan centered in the tumor volume using a 24-mm collimator. The results of this standard arc set are shown in Figures 6.17A–D. Note the significant amount of normal brain tissue around the tumor volume and the orientation of the lower isodose shells over the brain stem.

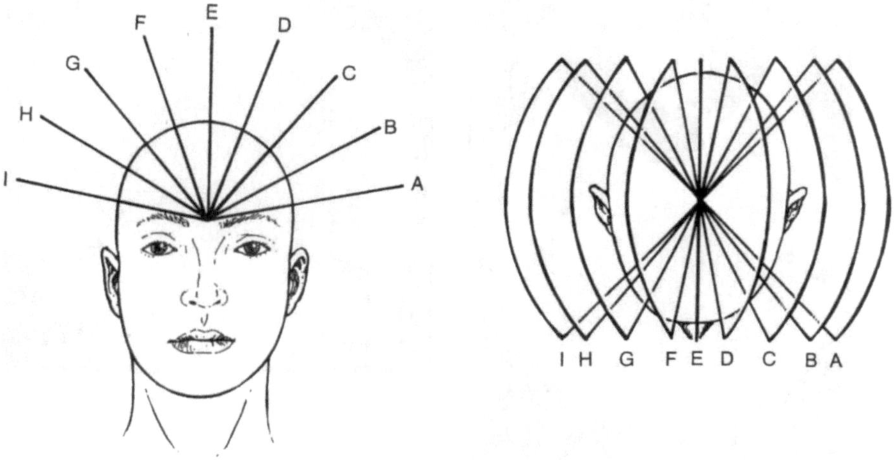

(A)

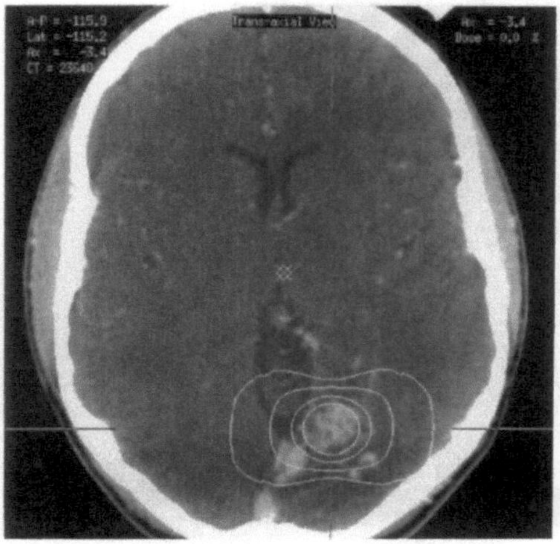

(B)

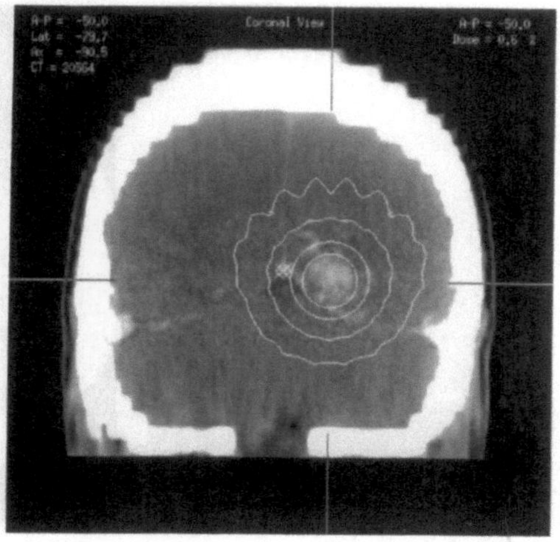

(C)

Figure 6.16. Spherical AVM. (A) Schematic view of our standard nine-arc treatment plan from the coronal and axial view. The arcs are lettered for comparison with subsequent examples (see Fig. 6.6). The dose distribution in the axial plane (B), the coronal plane (C), and the sagittal plane (D). The innermost isodose line is the 80% line followed by the 40%, 16%, and 8% isodose lines centripetally. Note that the image resolution in the coronal and sagittal planes above and below the target is degraded because of the larger (5 mm) CT scan slice thickness in these regions (see Figs. 4.8 through 4.10).

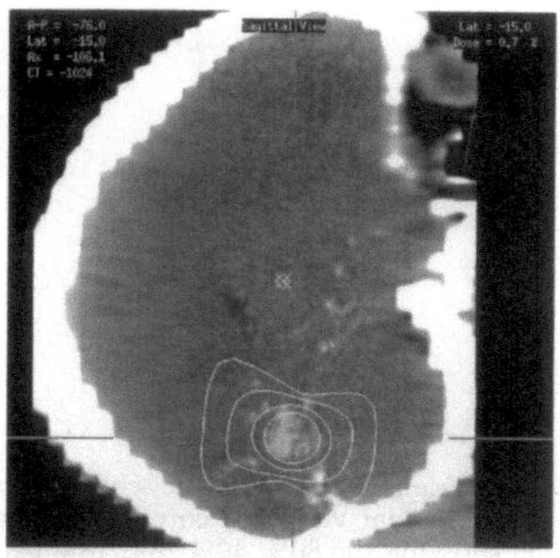

(D)

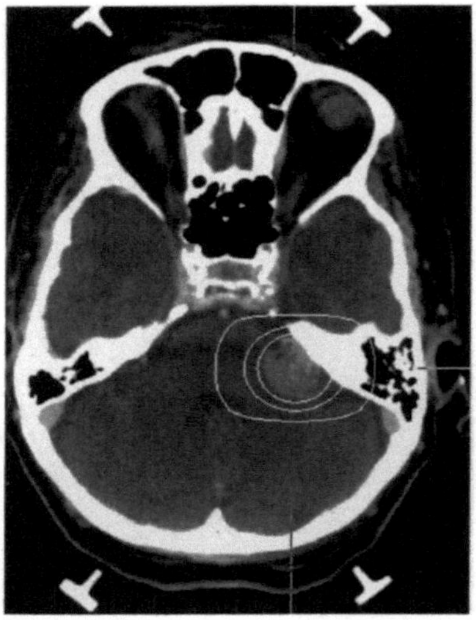

(A)

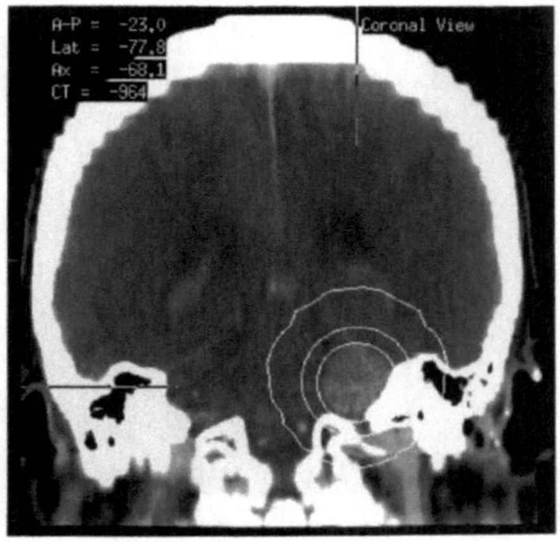

(B)

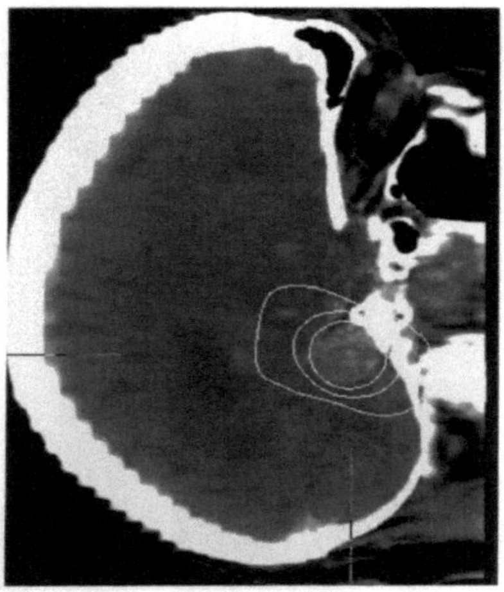

(C)

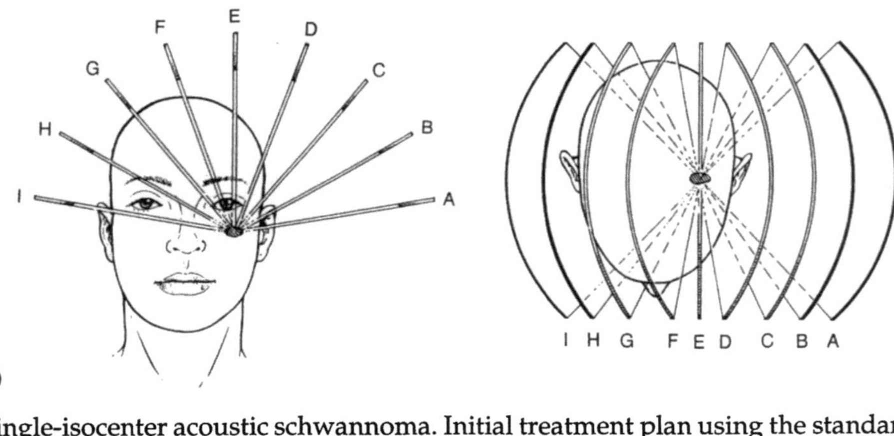

(D)

Figure 6.17. Single-isocenter acoustic schwannoma. Initial treatment plan using the standard nine-arc set. (A) Axial view, (B) coronal view, and (C) sagittal view of the resultant dose distribution showing the 80%, 40%, and 16% isodose lines. (D) Schematic of standard set in coronal and axial views. Elimination of the most horizontal arcs leads to the distribution shown in (E) axial, (F) coronal, and (G) sagittal views. The change is shown schematically in (H) on the coronal and axial views. Note that arcs A, B, H, and I have been eliminated (weighted 0) in the new plan. This achieves the primary treatment planning objective of dose gradient optimization for this lesion, which is located immediately lateral to the brain stem.

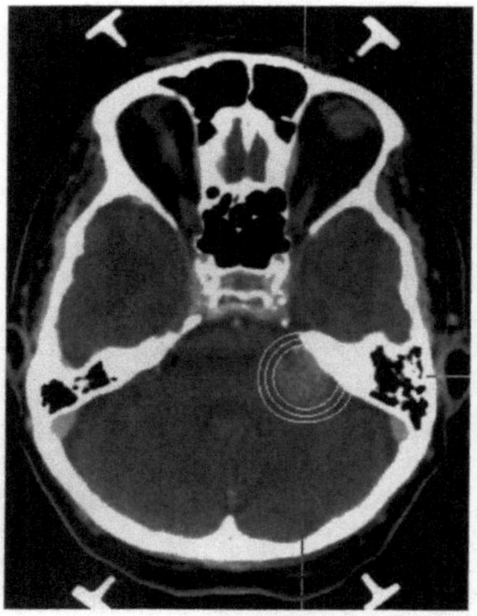

(E)

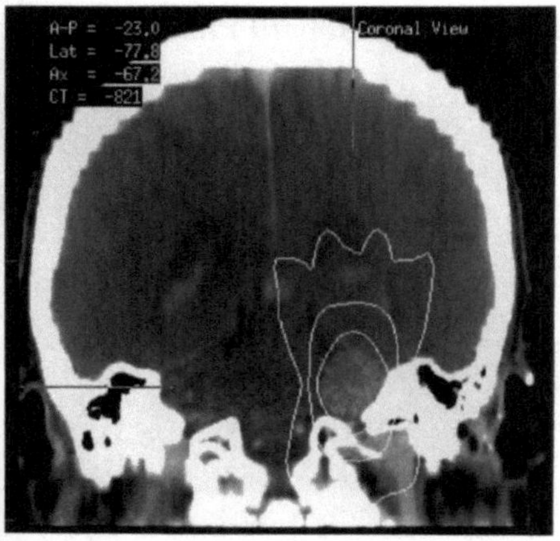

(F)

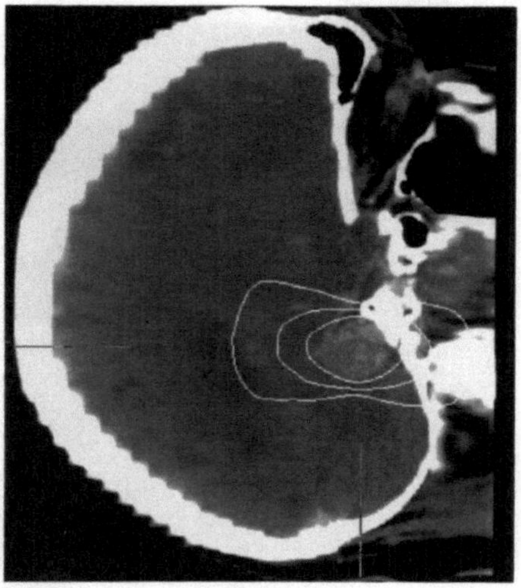

(G)

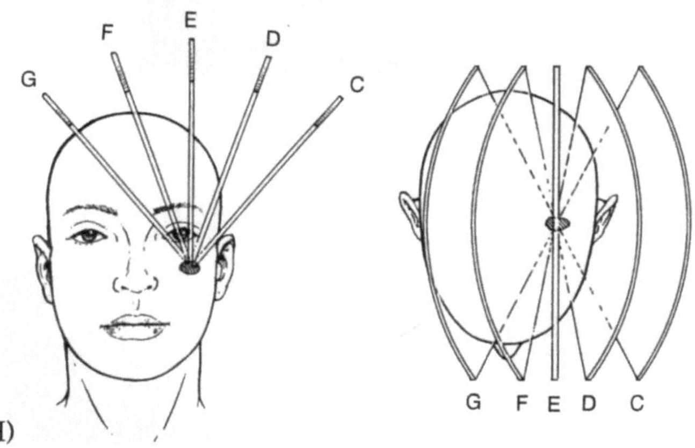

(H)

Figure 6.17. *Continued* A final modification involves decreasing the size of the most horizontal arcs (C and G). The resultant change in the dose distribution is shown in figures (I) axial, (J) coronal, and (K) sagittal planes. Note that there is no decrement in the dose gradient, yet the rim of normal tissue superiorly and inferiorly is diminished (i.e., the plan is more conformal) by decreasing the height of the distribution. This change is shown schematically in (L) in both coronal and axial planes, with decrease in size of the collimator represented by thinner lines in the arcs changed.

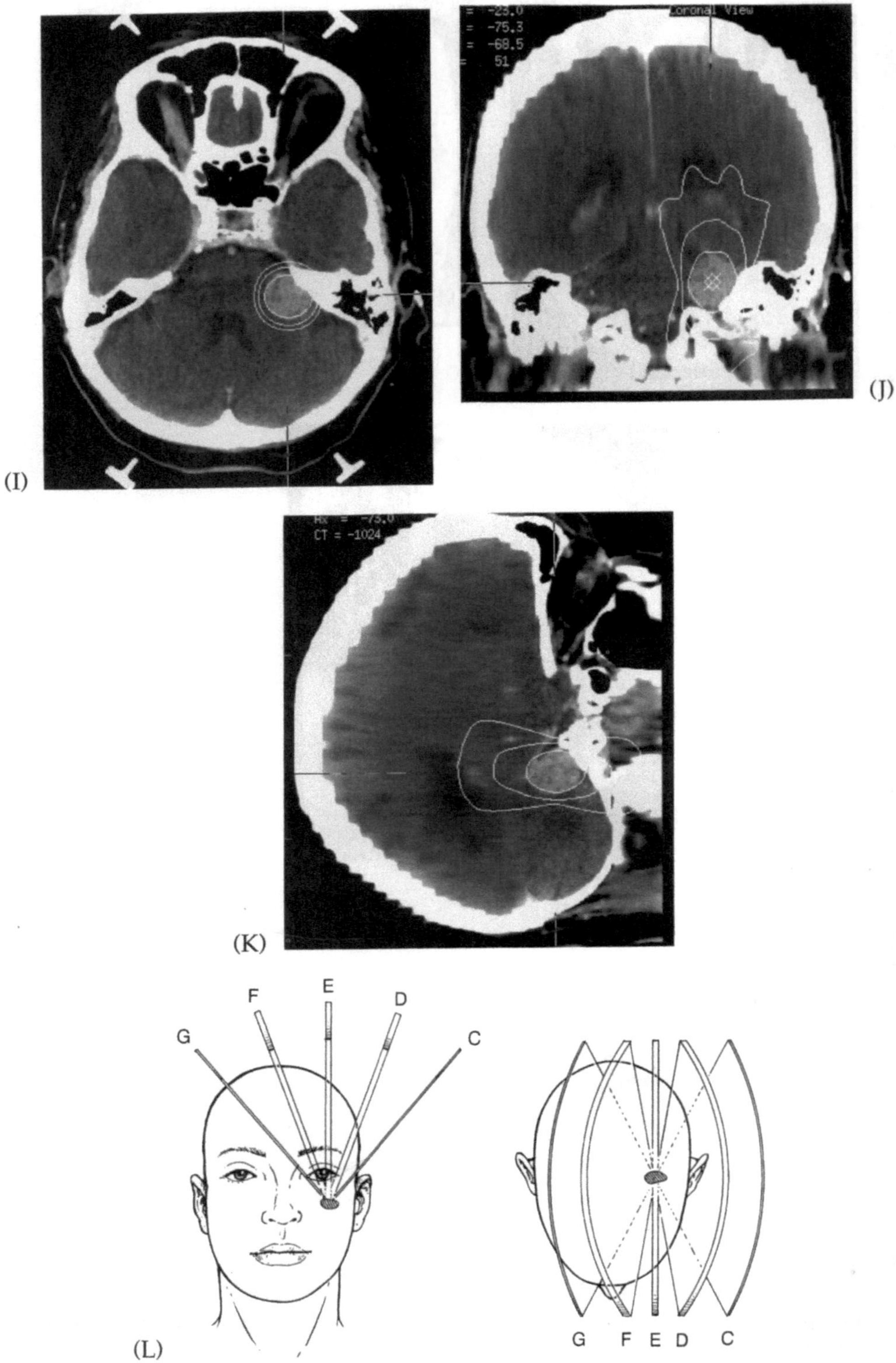

Figure 6.17. *Continued*

The initial change we make is to use the arc elimination tool. We eliminate the most horizontally oriented arcs, which gives us the distribution shown in Figures 6.17E–H. This changes the distribution of the dose gradients and illustrates the Jell-O® principle clearly. The distribution becomes elongated in the superior–inferior direction, and more narrow in the medial–lateral dimensions. This increases (i.e., makes steeper) the dose gradient in the lateral and medial directions, thereby decreasing the dose to the brain stem, where the critical neurologic tissue is located. It also decreases the dose gradient in the superior–inferior direction, where there is less critical tissue. Note that in using this distribution with 24-mm collimators there is a rim of normal tissue that remains.

In order to improve this further (i.e., increase conformality) we use the differential collimator arc-weighting tool and decrease the size of the two most horizontally oriented arcs. This differential reduction in collimator size reduces the height of the dose distribution and further improves conformality. The final plan with arc orientations is shown in Figures 6.17I–L.

3. Mildly Ellipsoidal Metastasis

This case illustrates the isolated effect of changing the collimator sizes on selected arcs to change the shape of the isodose distribution in a more subtle fashion than with arc elimination. Figures 6.18A–C show a standard nine-arc plan that encompasses the tumor volume. The collimator size is 24 mm for all arcs, and the isodose lines are the 80%, 40%, and 16% in centripetal sequence. Note the significant rim of normal brain shown in the axial scan compared with the relatively good fit in the coronal scan. The collimator size in the five vertically oriented arcs is therefore reduced to 20 mm because the most vertical arcs control the width of the distribution. The four more horizontal arcs are maintained at 24 mm because they control the height of the distribution. Applying this change results in the dose distribution shown in Figures 6.18D–F. Note that the rim of normal tissue in the prescription volume is reduced, as are the volumes within the lower isodose lines. This illustrates a case where no critical neurologic structure is present and therefore maximal conformality is pursued using an arc-weighting tool. Assiduous pursuit of optimal plans for such apparently simple geometry minimizes complications and allows the dose to be maximized to the target because less normal brain is treated.

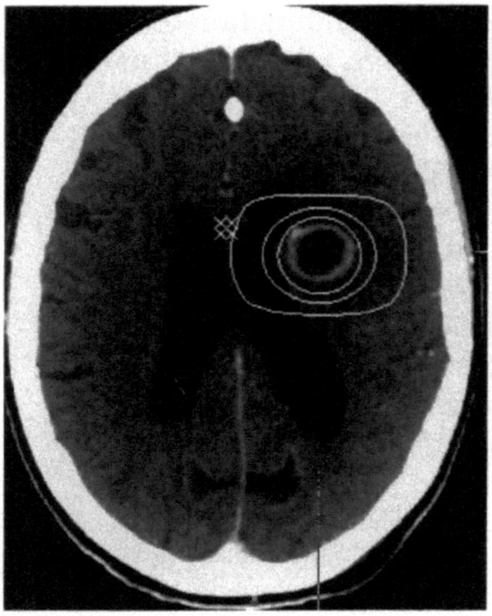

(A)

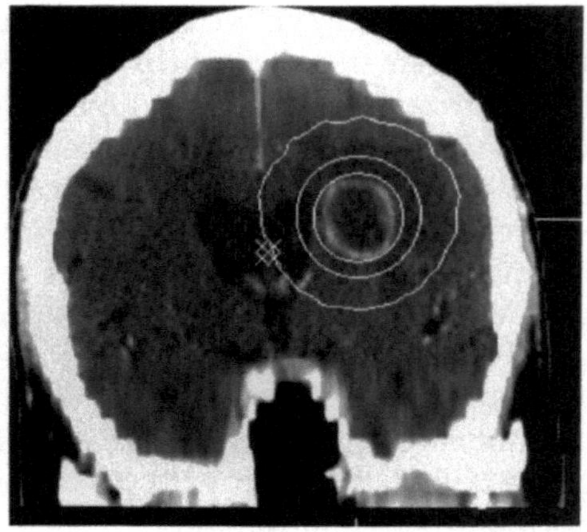

(B)

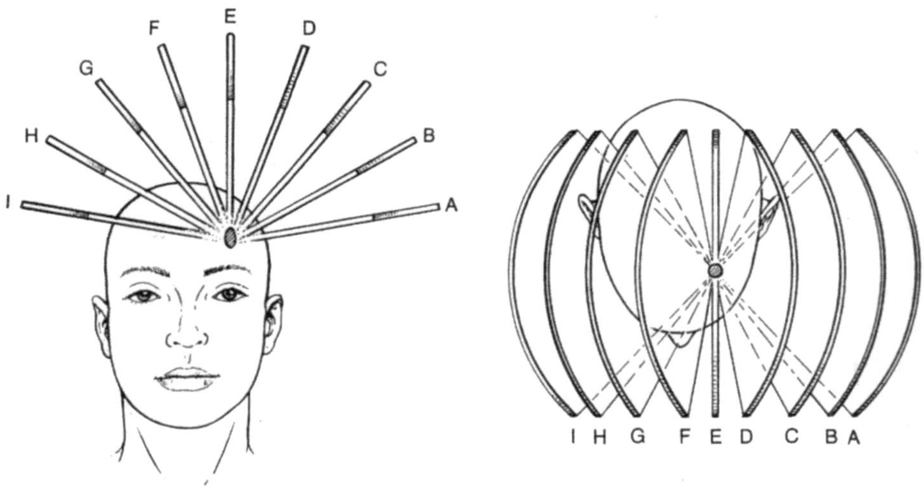

(C)

Figure 6.18. Mildly ellipsoidal. The smallest collimator that encompasses the target using the standard nine-arc plan is shown in the (A) axial and (B) coronal planes. The schematic of this is shown in (C). Because the primary objective in treatment planning for this lesion (distant from critical neurologic structures) is conformality, and the lesion is mildly ellipsoidal in the superior–inferior direction in the coronal plane, a change in collimator sizes is indicated using the algorithm. Decrease in collimator size (from 24 mm to 20 mm) for the five vertically oriented arcs C, D, E, F, and G results in the dose distribution shown in (D) and (E). This change is shown schematically in (F) by the thinner arcs in these vertical arcs (C–G). Note that the normal tissue included in the prescription line (80%) is decreased in both the axial and coronal planes.

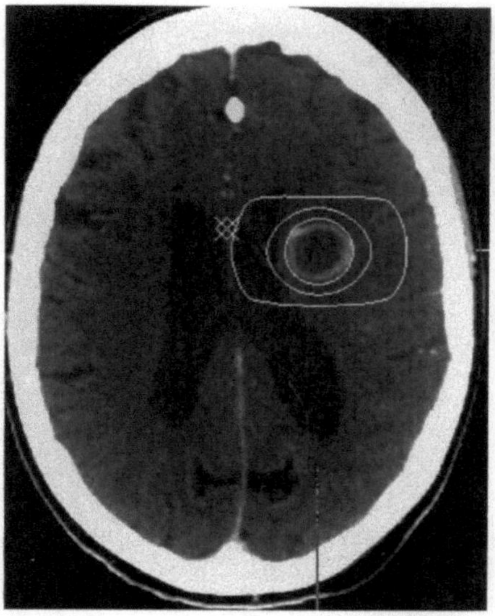

(D)

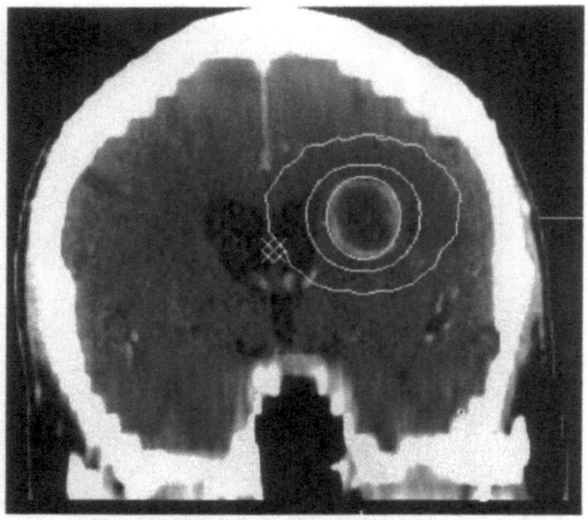

(E)

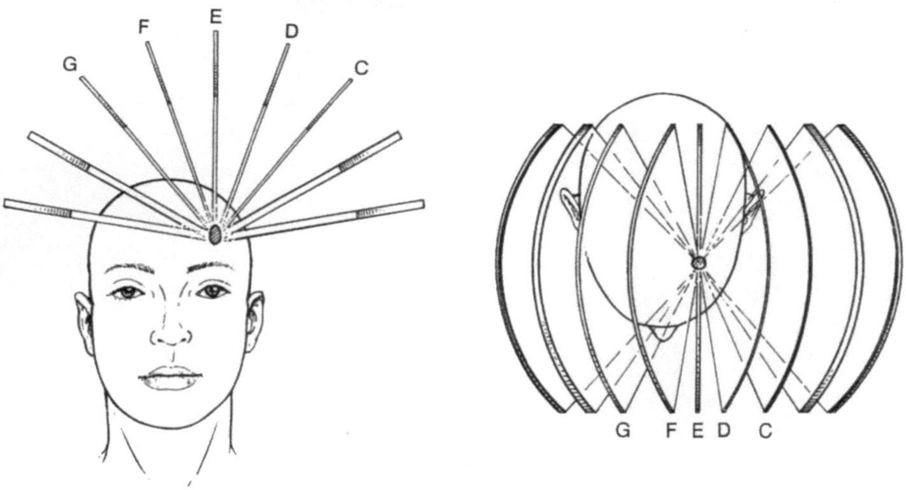

(F)

Figure 6.18. *Continued*

4. Ellipsoidal: Principal Axis Coronal Plane—Arteriovenous Malformation

This patient initially had a hemorrhage. The case illustrates the effect of arc elimination in obtaining a more optimal dose distribution for an arteriovenous malformation that is ellipsoidal with the principal axis horizontal in the coronal plane. The minimally encompassing standard nine-arc plan is shown in Figures 6.19A–D. Isodose lines are again the 80%, 40%, and 16%. The plan uses a 35-mm collimator, is poorly conformal, treats a large amount of normal brain superior and inferior to the lesion, and is inadequate in the medial–lateral dimensions. The three most vertically oriented arcs (Figs. 6.19D–F), therefore, are first

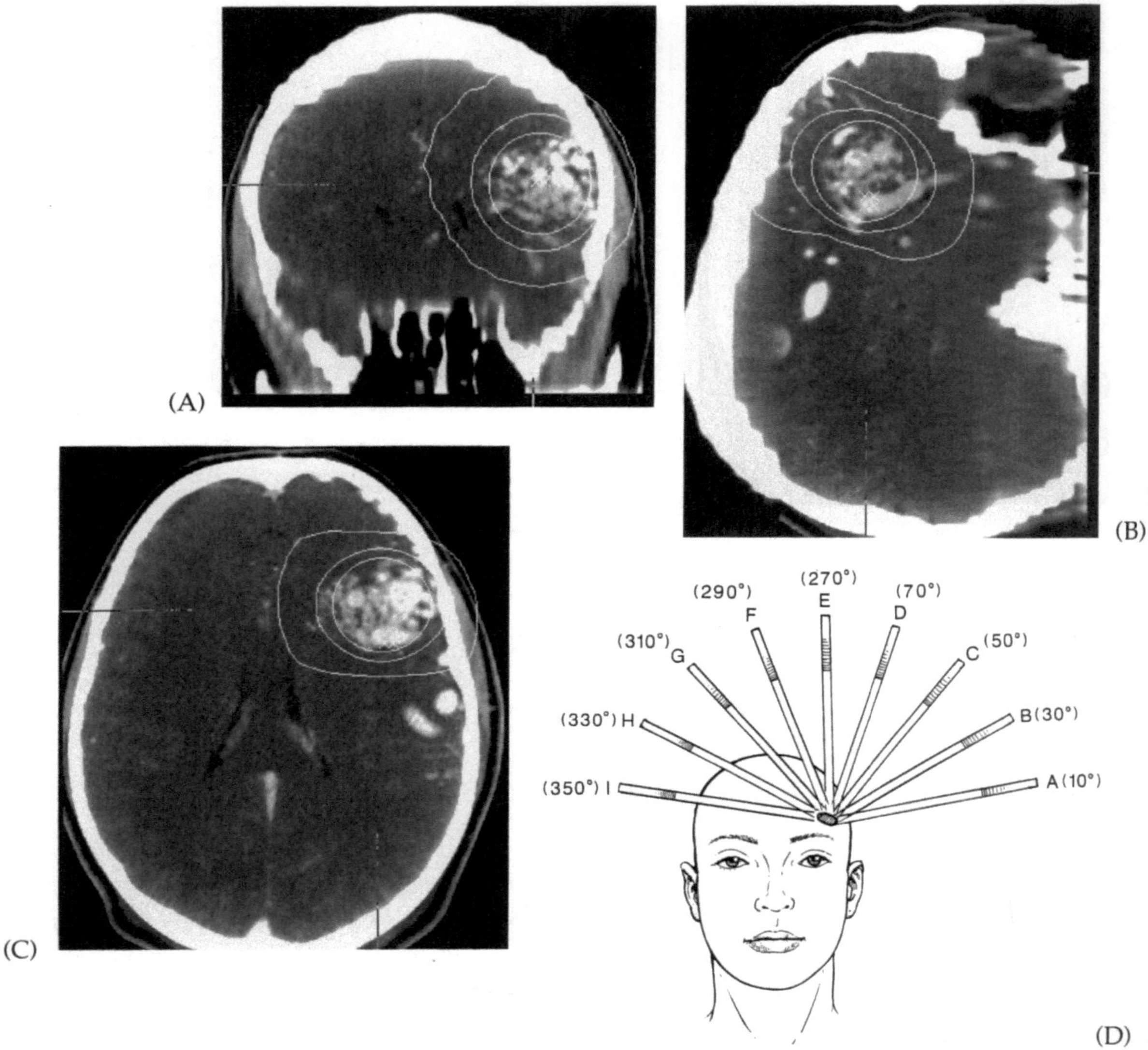

Figure 6.19. Ellipsoidal. Principal axis in the coronal plane. The minimally encompassing nine-arc plan is shown in the (A) coronal, (B) sagittal, and (C) axial planes. The schematic of this starting point with arc angles is shown in (D). Adjustment is first made by eliminating the most vertical arcs lettered D, E, and F. The lesion, however, is not strictly horizontal; therefore, arcs B and C are also eliminated and replaced by a new arc labeled X at 0 degrees that approximates the principal axis of the lesion. The result of these changes is shown in the (E) coronal, (F) sagittal, and (G) axial planes, and in (H) the schematic illustration of these changes. The final treatment-planning change involves decreasing the collimator sizes in the most horizontally oriented arcs from 35 mm to 28 mm (arcs H, I, A, and X). The result of this change is shown in the (I) coronal, (J) sagittal, and (K) axial planes, as well as schematically in (L).

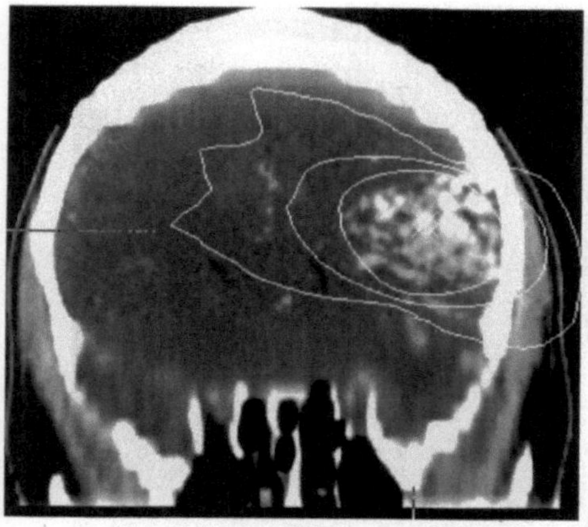

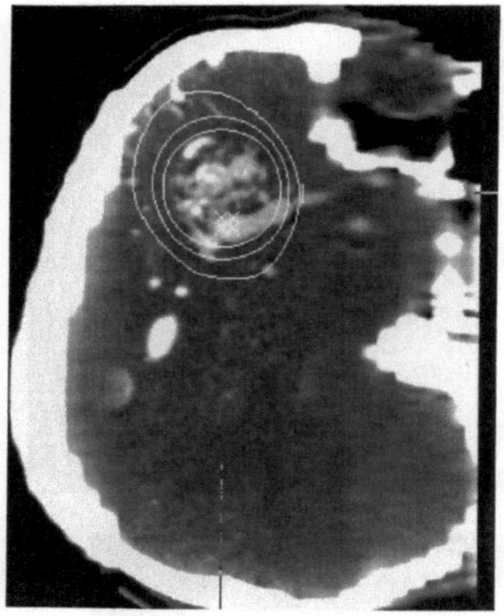

(F)

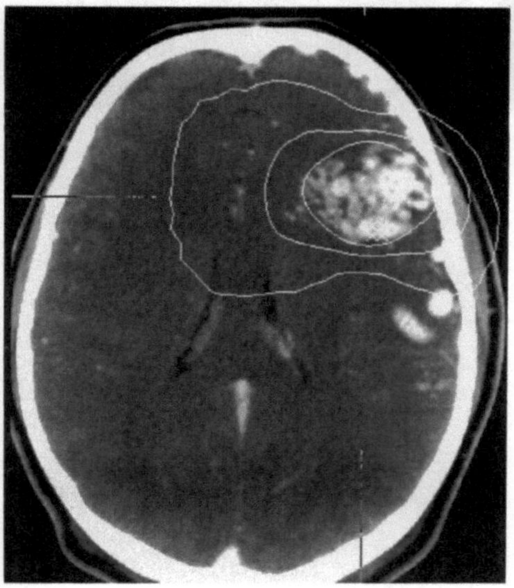

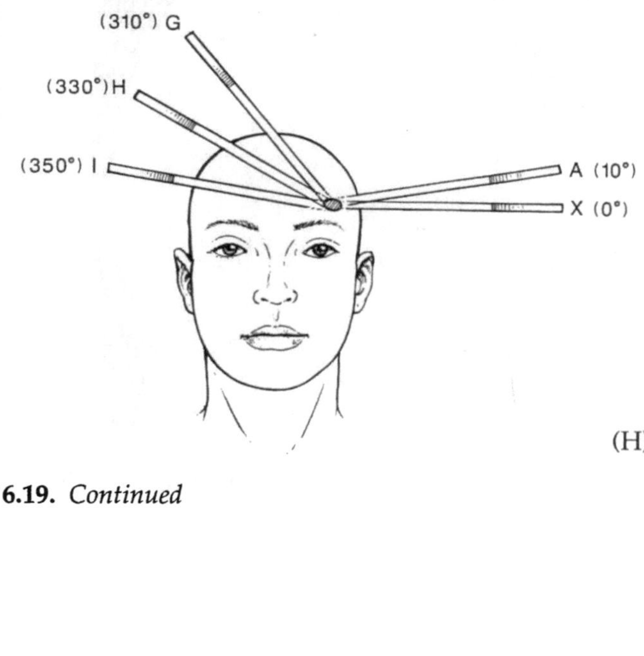

(H)

Figure 6.19. *Continued*

eliminated to achieve a horizontally ellipsoidal distribution in the principal axis of the lesion. This lesion, however, is not truly horizontal in the coronal plane and some additional changes are therefore needed to more appropriately tilt the distribution in the principal axis of the ellipsoid. To approximate this principal axis, arcs C and B are eliminated next, and an arc is added at 0 degrees (labeled *x*). This is shown in Figures 6.19E–H. This distribution is now appropriately tilted and adequate in the medial–lateral dimension, but it still treats a large volume of normal tissue superior and inferior to the lesion because the

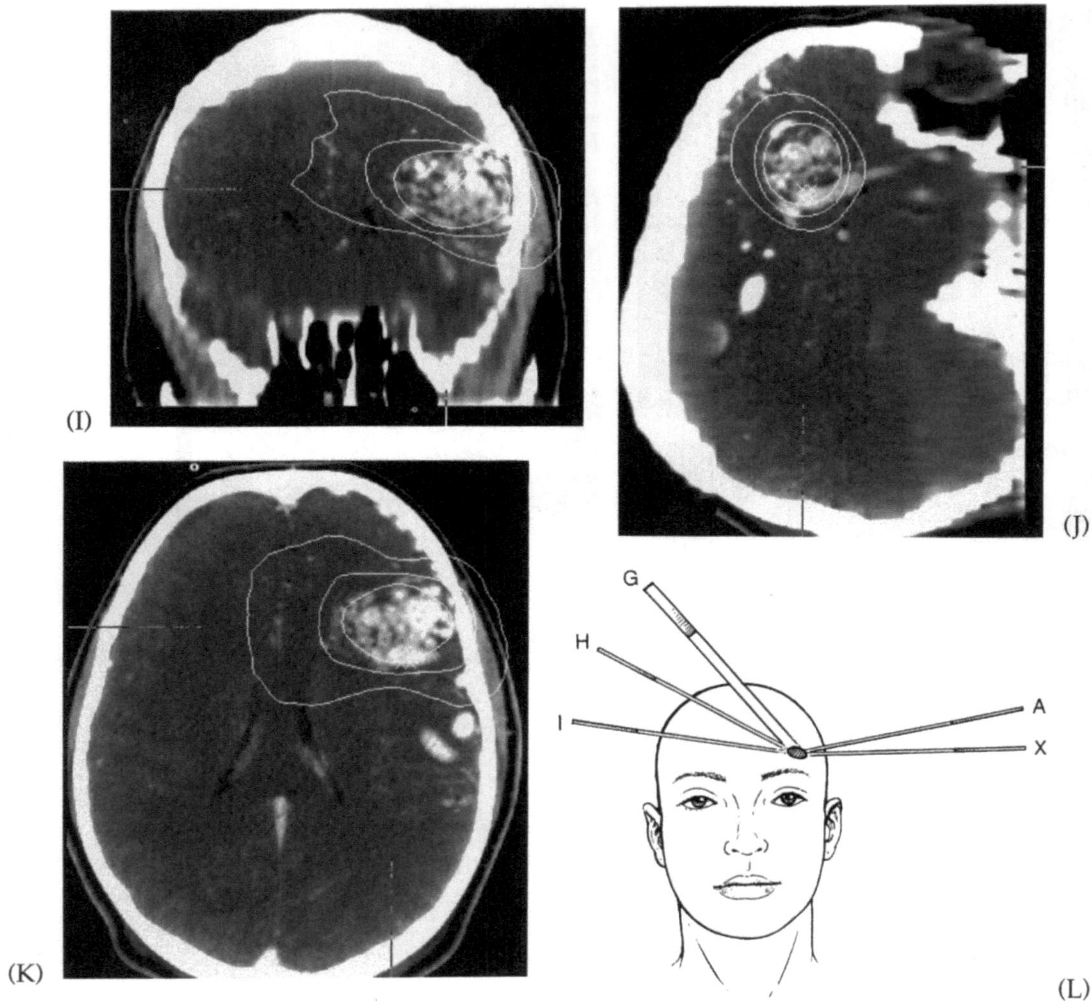

Figure 6.19. *Continued*

collimator size of 35 mm has not been changed. The four most horizontally oriented collimators are reduced to 28 mm in Figures 6.19I–L because the height of the distribution needs to be decreased. The conformality is dramatically improved.

5. Ellipsoidal in the Coronal Plane: Critical Structure Superior and Anterior to the Lesion—Pituitary Adenoma

This patient presented with Cushing's disease and had a microadenoma evident radiographically. She refused surgical intervention and was therefore referred for radiosurgery. The initial step in the algorithm asks whether there are any adjacent critical neural structures, and in this case the optic chiasm is noted superior to the lesion and the optic nerves are anterior–superior to the lesion. Figures 6.20A–D illustrate the results of the standard nine-arc distribution. The 80%, 40%, and 16% isodose lines are shown. Note that the vertical arcs D, E, and F pass directly through the optic chiasm region (Fig. 6.20D), which exposes the optic chiasm to approximately 20% of the prescription dose (Figures 6.20B,C).

The first step in applying the algorithm, therefore, is to delete (weight 0) the three vertical arcs (D, E, F). This results in the distributions shown in Figures 6.20E–G that is illustrated schematically in Figure 6.20H. Although this is a significant improvement, the lesion is also close to the optic nerve, which sits slightly anterior and superior to the lesion. The most anterior 30 degrees of each arc, therefore, is removed to reduce the arc span over the optic nerves. This results in the distributions shown in Figures 6.20I–K that is demonstrated schematically in 6.20L. Note that this change leads to a further elongation of the lower isodose lines laterally with reduction in the AP dimension. This plan accomplishes the primary goal of reducing the spread of the distribution both superiorly and anteriorly. The plan remains conformal to the microadenoma. The lower isodose lines are elongated toward the more radioresistant cavernous sinus structures.

6. Two-Isocenter Acoustic Schwannoma

After reviewing the imaging it is clear that some lesions are not spherical or ellipsoidal, and therefore require multiple isocenter plans. Our first endeavor is to identify the plane of the lesion that best displays the principal axis of the lesion. This is usually in the axial plane for acoustic schwannomas. The plane used for planning is shown in Figure 6.21A. Note that this is an MR image that is more commonly used in our treatment planning of acoustic schwannomas because the lesion visualization is better than it is with CT.

After identifying this plane, we measure the length of the principal axis of the lesion (i.e., by using a measuring tool) and attempt to evenly distribute the isocenters through the axis. The first isocenter is placed as shown in Figure 6.21B. A collimator size is chosen that will minimally cover the medial portion of the lesion. The medial collimator size in this case is 14 mm. Note that the five-arc plan that maximizes the medial gradient is still used here (Fig. 6.21C and Case Example 2). Our next step is to position the second isocenter, as shown in Figures 6.21D,E. A 10-mm collimator adequately covers the intracanalicular component of

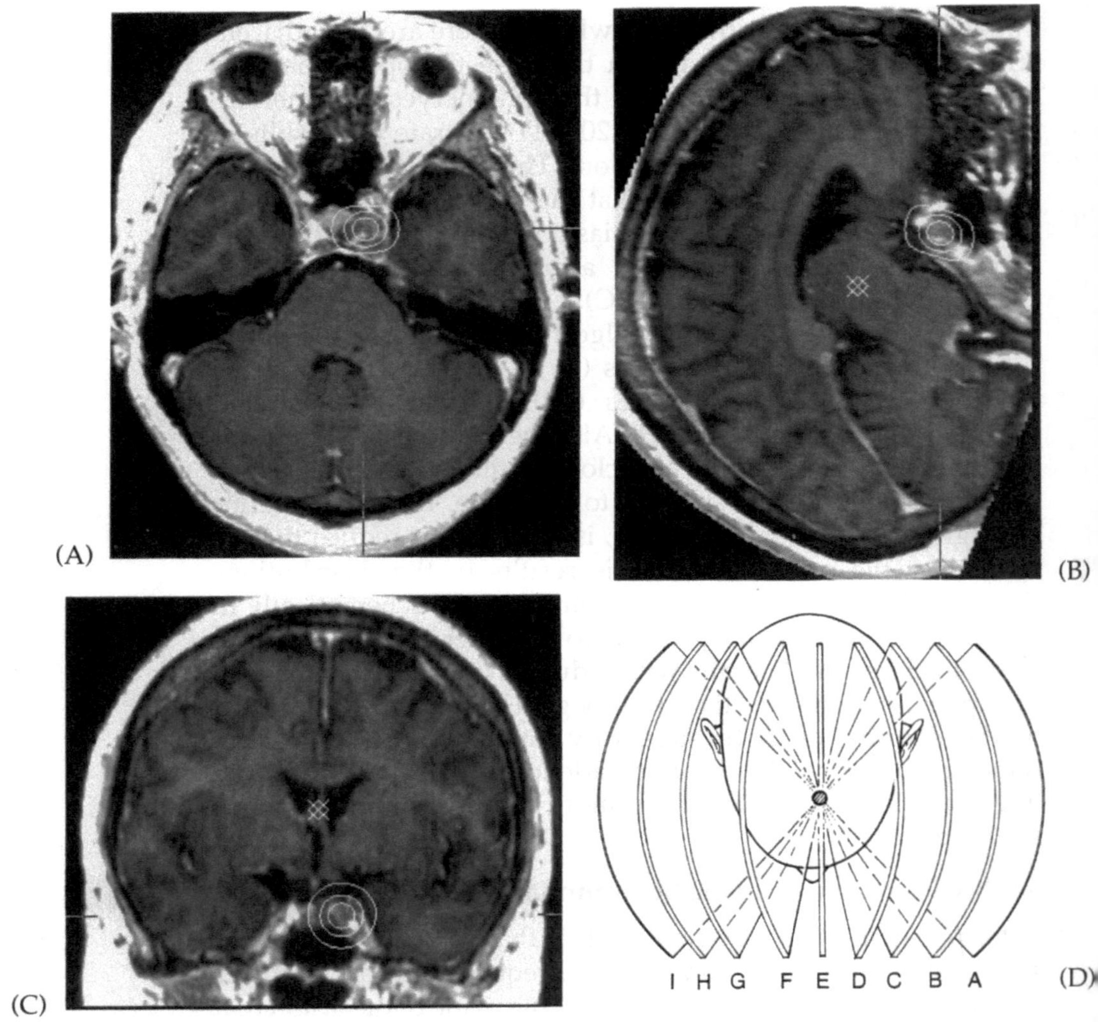

Figure 6.20. Ellipsoidal in the coronal plane. Critical structure superior and anterior to the lesion. The standard nine-arc plan that minimally encompasses the pituitary adenoma uses a 10-mm collimator and is shown in the (A) axial, (B) sagittal, and (C) coronal views. The schematic of this starting point is (D). Eliminating lettered arcs D, E, and F leads to the dose distributions shown in (E), (F), and (G). These changes are illustrated in (H). This significantly reduces the superior spread of the dose distribution and avoids the critical neural structure, the optic chiasm. To improve the distribution, the anterior 30 degrees are removed from each of the remaining horizontally oriented arcs. This reduces the dose of radiation through the optic nerves located anterior and superior to the lesion. The results of this change are shown in (I), (J), and (K), as well as schematically in (L).

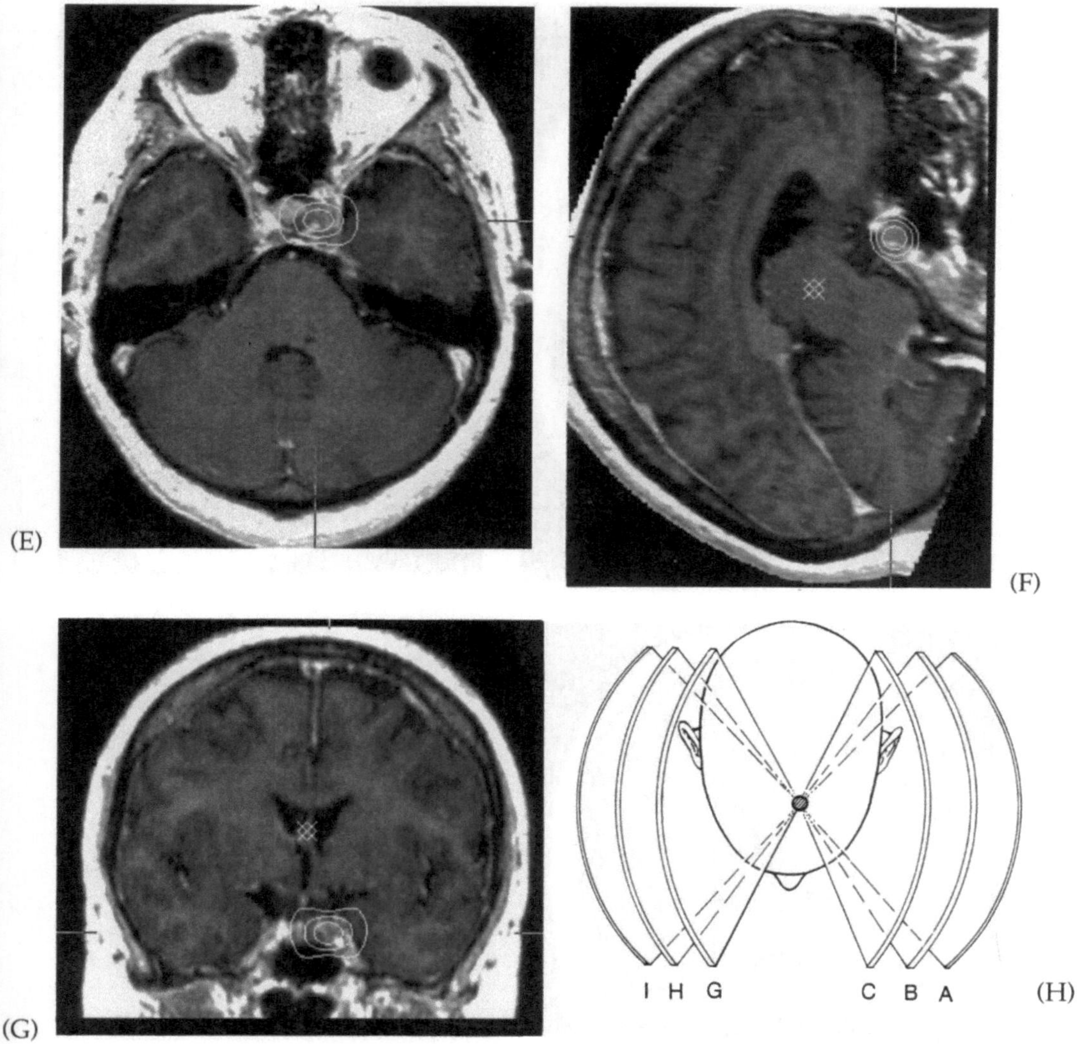

Figure 6.20. *Continued*

the lesion in the more lateral isocenter. Because the more lateral isocenter is relatively distant from the brain stem and because we desire significant interaction with the first isocenter, the arcs used for the second isocenter are generally evenly spaced (Fig. 6.21E), as opposed to the vertically oriented arcs used to maximize the medial gradient in the first isocenter (Fig. 6.21C). We then change from our standard isodose designations to make the most central isodose 70% of maximum, and subsequent isodose lines 35%, and then 14%. (Note that prescribing to slightly lower isodose lines in multiple isocenter plans is usually required to obtain useful dose distributions.) We then assess the

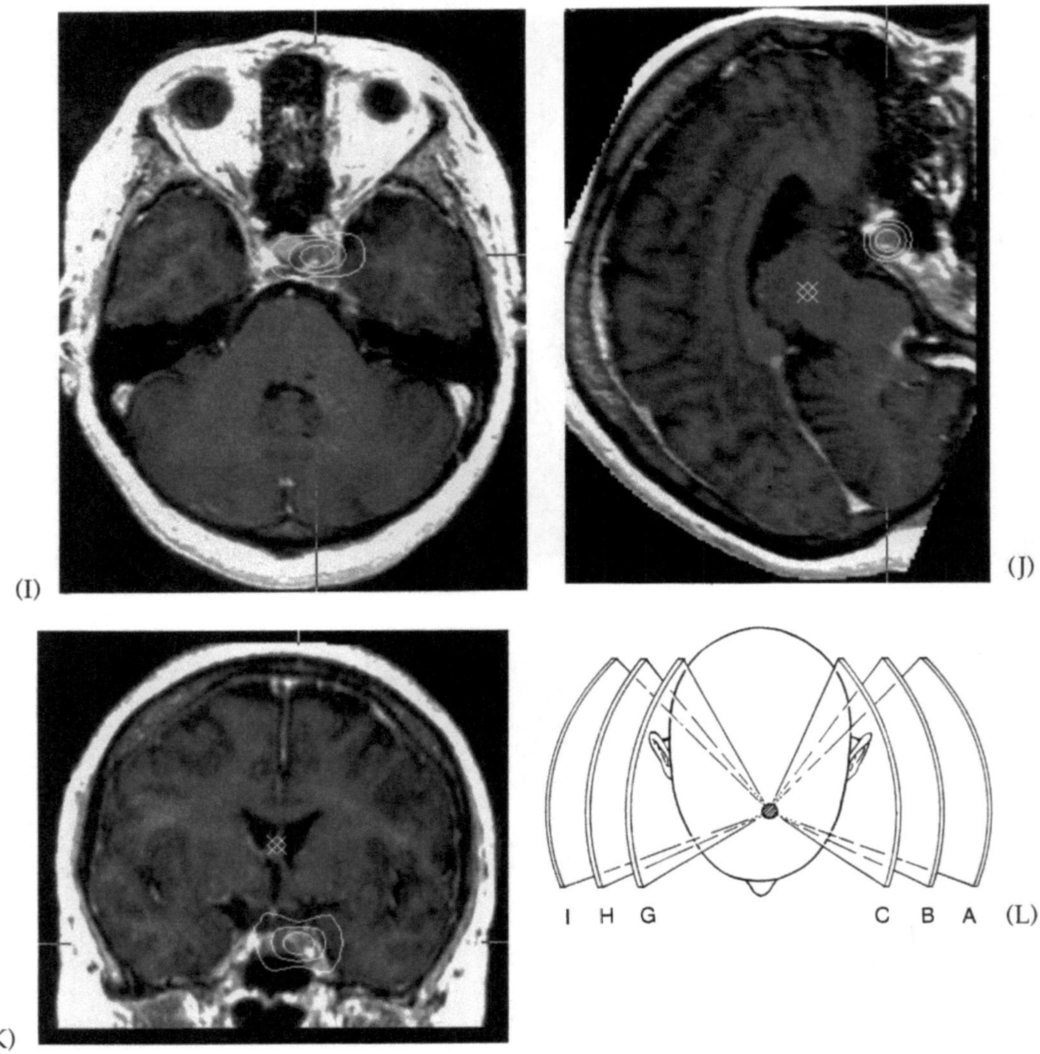

Figure 6.20. *Continued*

optimal spacing for a 14-mm and 10-mm collimator using our spacing tool, as shown in Figure 6.21F. (We have found that this tool gives us the optimal spacing for any two collimators and is extremely useful for obtaining ideal spacing.)

Once we are aware of the ideal spacing as opposed to our attempted spacing, we can move the isocenters so that the lesion is optimally encompassed and normal tissue is maximally spared. The final plan is shown in Figures 6.21G,H. This planning approach is frequently used for small acoustic schwannomas and illustrates many of the principles of multiple-isocenter treatment planning. The resulting conformality of the plan is excellent and it maintains the medial dose gradient adjacent to the brain stem. It is important to realize that the spacing of the isocenters may need to be slightly altered from the ideal and that this process is iterative. A small change in spacing often has dramatic effects on the distribution, and several moves may be necessary before achieving the desired effect. The computing necessary for multiple isocenter plans is greater and each move requires complete recalculation of the doses by the computer. We have found rapid computational ability to be nearly essential in this process.

7. Irregular: Principal Axis in the Axial Plane—Metastasis

This case illustrates that an irregular lesion with its principal axis in the axial plane requires multiple isocenters. The placement of the first two isocenters is shown in Figures 6.22A,B. The principles used in their placement are identical to those used in Case Example 6. The spacing tool is used again to determine the optimal spacing of the two isocenters, as shown in Figure 6.22C. The resulting plan displaying the 70%, 35%, and 14% isodose lines is shown in Figure 6.22D.

8. Multiple Isocenter Arteriovenous Malformation

This case illustrates an example of a more complex three-isocenter plan for an arteriovenous malformation. One of the most critical steps in identifying a treatment approach for a lesion such as this one is the ability to obtain an image in the plane of the principal axis of the lesion, or what is often called the *three-point view* (Fig. 6.13). For this case, neither the axial, coronal, nor sagittal planes give an adequate appreciation of the three-dimensional configuration of the lesion; therefore, the off-axis view through the three isocenters chosen is shown in Figure 6.23. Once this view is obtained, the isocenters are placed and manipulated as illustrated in Case Examples 6 and 7. The spacing tool is extremely valuable for ensuring that spacing of all isocenters is optimal. The complexity of the isocenter interactions increases with number, as does the need for faster computing times. Expertise and commitment to an optimal plan during the iterative process of designing complex plans is the cornerstone of outstanding treatment planning. The isodose lines represented in Figure 6.23 are the 70%, 35%, and 14%, respectively.

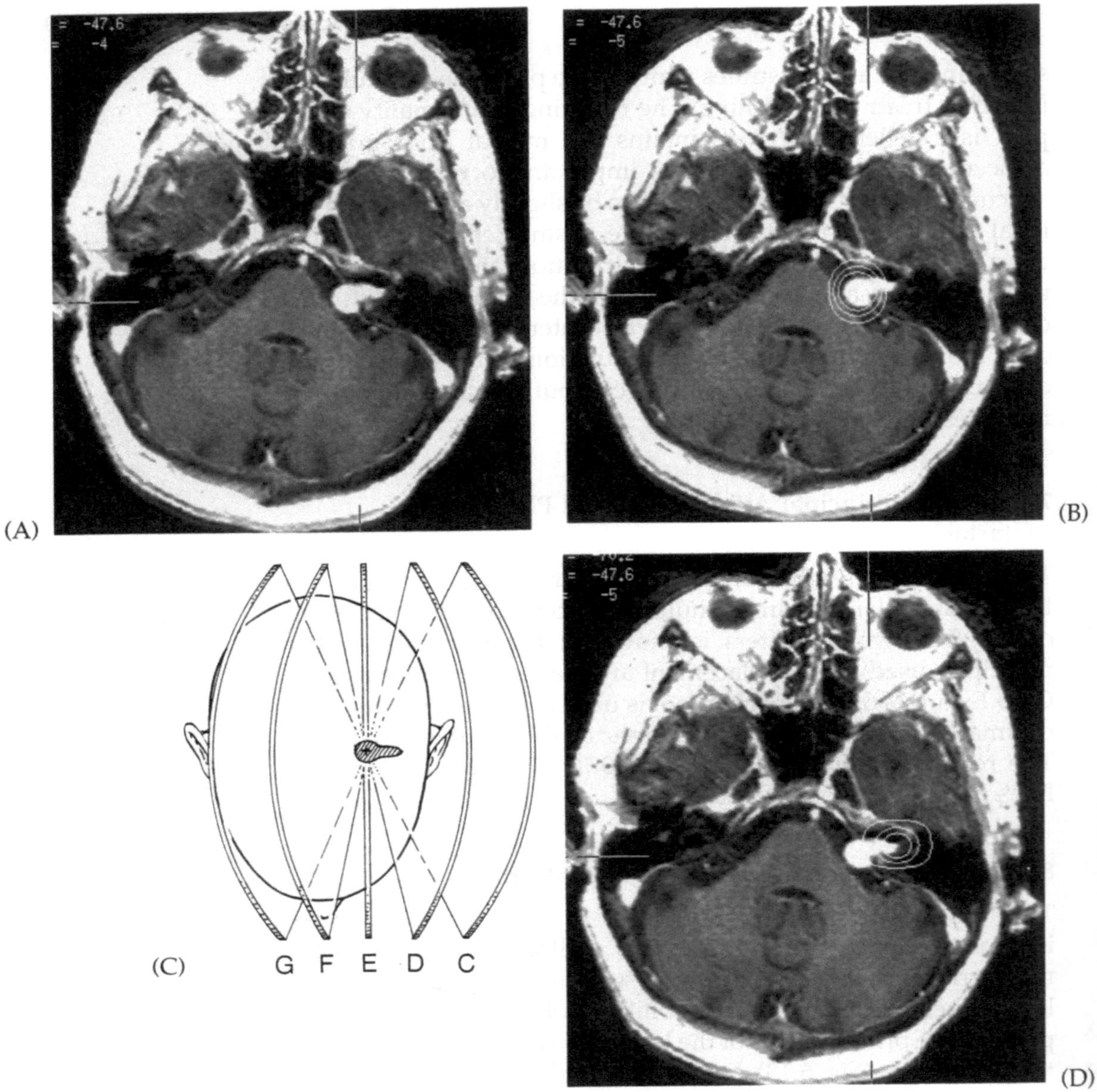

Figure 6.21. Two-isocenter acoustic schwannoma. (A) An MRI of an acoustic schwannoma best treated with a two-isocenter plan. Note that the lesion is small but has prominent intracranial and intracanalicular components. Use of MRI as opposed to CT scans allows adequate visualization of both of these components. Planning begins by identifying this plane and then distributing two isocenters with appropriately small collimators on the intracranial and intracanalicular components (e.g., 14 mm for the intracranial component and 10 mm for the intracanalicular component in this case). The dose distribution for the first (i.e., intracranial) isocenter is shown in (B) with the axial arc set typically used to protect the medial brain stem (C) (see Case Example #2). Isodose lines 80%, 40%, and 16% are shown. The second (i.e., intracanalicular) isocenter is then positioned and uses an equally distributed five-arc set (D) and (E). Next the isodose lines representd are changed to the 70%, 35%, and 14% isodose shells required by multiple isocenter planning. The spacing tool is then checked (F). The isocenters are then moved to approximate the ideal, and the composite distribution is reviewed. Optimal spacing is a starting point, but several iterative changes may be needed to achieve optimal spacing for a given case. The resulting conformal distribution is shown in the axial view (G), as well as schematically (H).

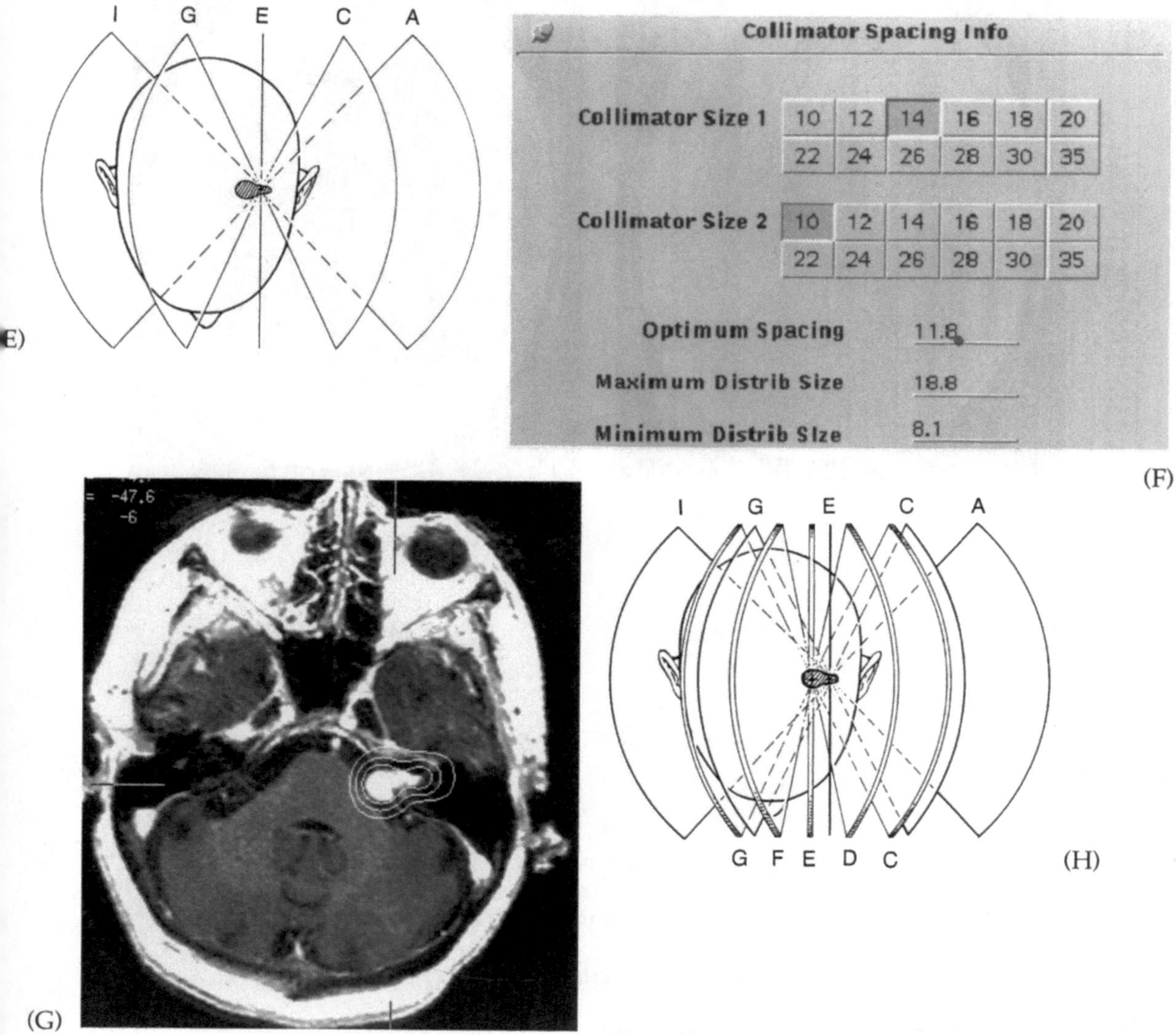

Figure 6.21. *Continued*

9. Critical Structure Anterior to the Lesion—Hemangioblastoma

This patient has Von Hippel Lindau syndrome and a solitary hemangioblastoma in the anterior fourth ventricle region immediately posterior to the brain stem. The lesion is quite spherical in shape; therefore, a standard nine-arc plan was initially chosen, as shown in Figures 6.24A–D. Note that the distribution is adequate for the lesion, but because the hemangioblastoma is immediately adjacent to the critical and radiosensitive brain stem, the primary goal of treatment

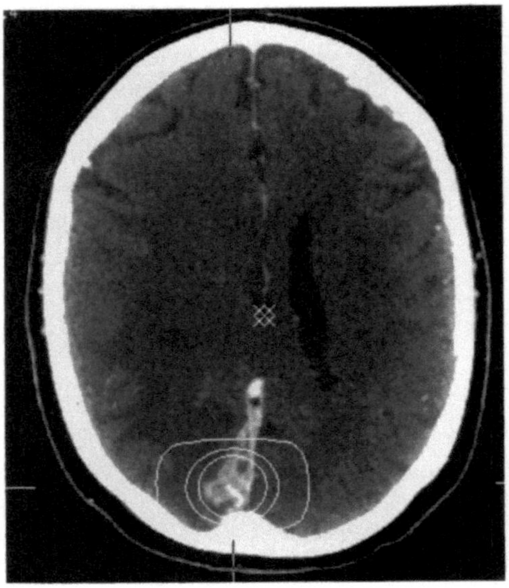

(A)

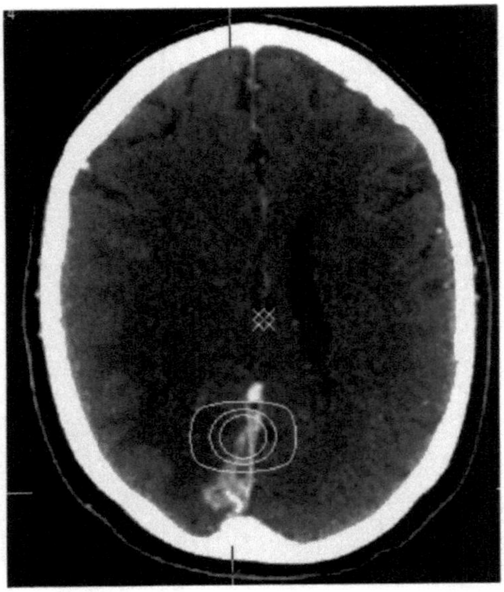

(B)

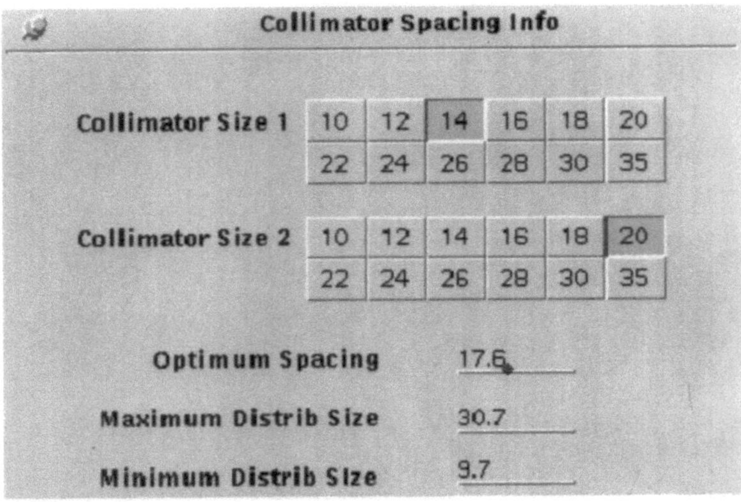
Collimator Spacing Info
Collimator Size 1
10 12 14 16 18 20
22 24 26 28 30 35
Collimator Size 2
10 12 14 16 18 20
22 24 26 28 30 35
Optimum Spacing 17.6
Maximum Distrib Size 30.7
Minimum Distrib Size 9.7

(C)

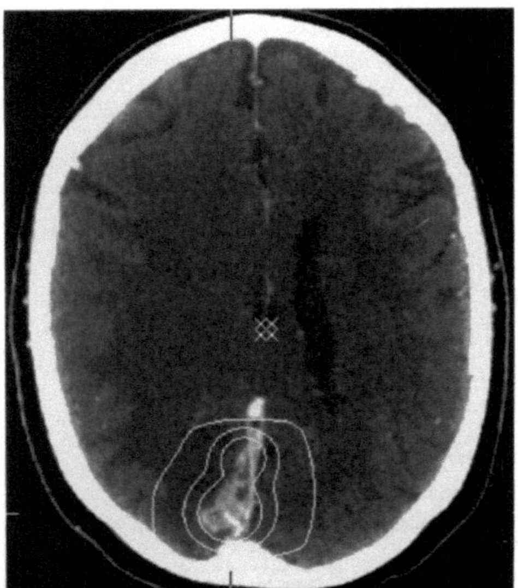

(D)

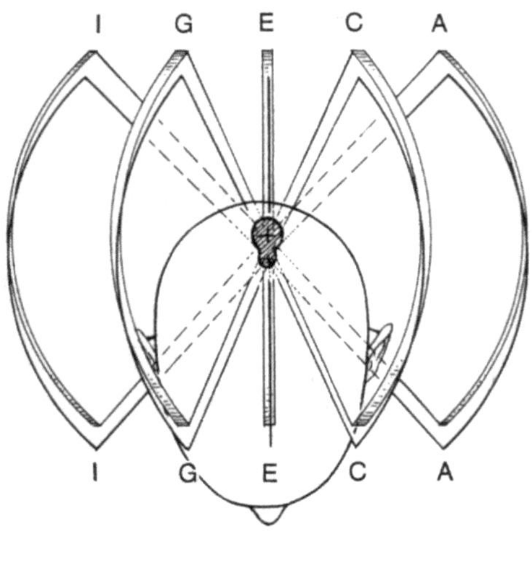
I G E C A
I G E C A

(E)

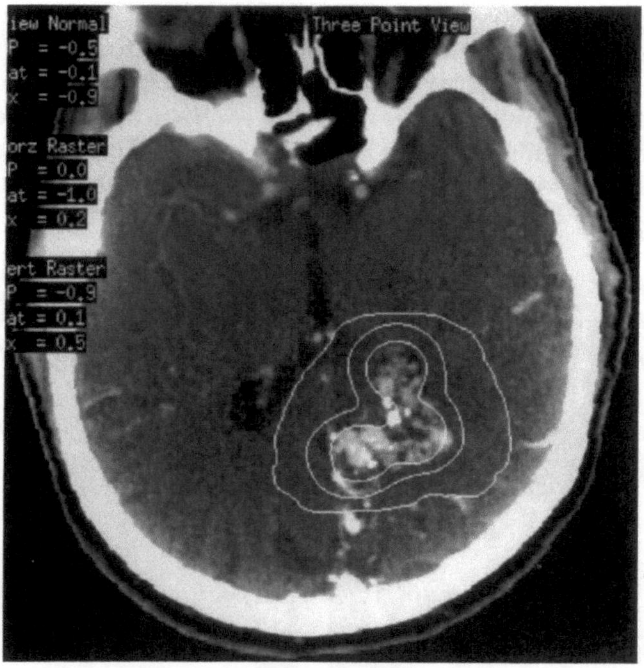

Figure 6.23. Multiple isocenter. This three-point view of the arteriovenous malformation allows treatment planning to occur in unique planes based on the axis and configuration of the lesion. In this case a three-isocenter plan results in the distribution shown. The 70%, 35%, and 14% isodose lines are displayed.

Figure 6.22. Irregular. Principal axis in the axial plane. This irregular lesion requires multiple isocenters because it is nonellipsoidal and has its principal axis of elongation in the AP direction in the axial plane. With this plan, the lesion is measured, then the larger posterior isocenter position is selected. A 20-mm collimator is selected and the resulting dose distribution with the 80%, 40%, and 16% isodose lines is shown in (A). Note that in multiple-isocenter plans not adjacent to critical structures, the equally weighted five-arc plan (see Fig. 6.21E) becomes the standard arc set for each isocenter. The position of the second, more anterior isocenter is then selected, and a 14-mm collimator is used, which results in the distribution shown in (B). The spacing tool is selected (C), and isocenters are moved accordingly after changing the dose prescription lines to the 70%, 35%, and 14% lines, respectively. The final plan is shown in the axial plane in (D) and in a schematic representation (E). Note the relationships of the arcs in the first isocenter (i.e., thicker lines) to the arcs of the second isocenter (i.e., thinner lines).

planning will be to optimize the dose gradient for the anterior critical structure. The best approach for this is to reduce the arc span of all arcs anteriorly, so that the area going through the brain stem is minimized. By reducing each arc by 20 degrees, we generate the plan shown in Figures 6.24E–H. The plan remains optimally conformal and now reduces the volume of anterior

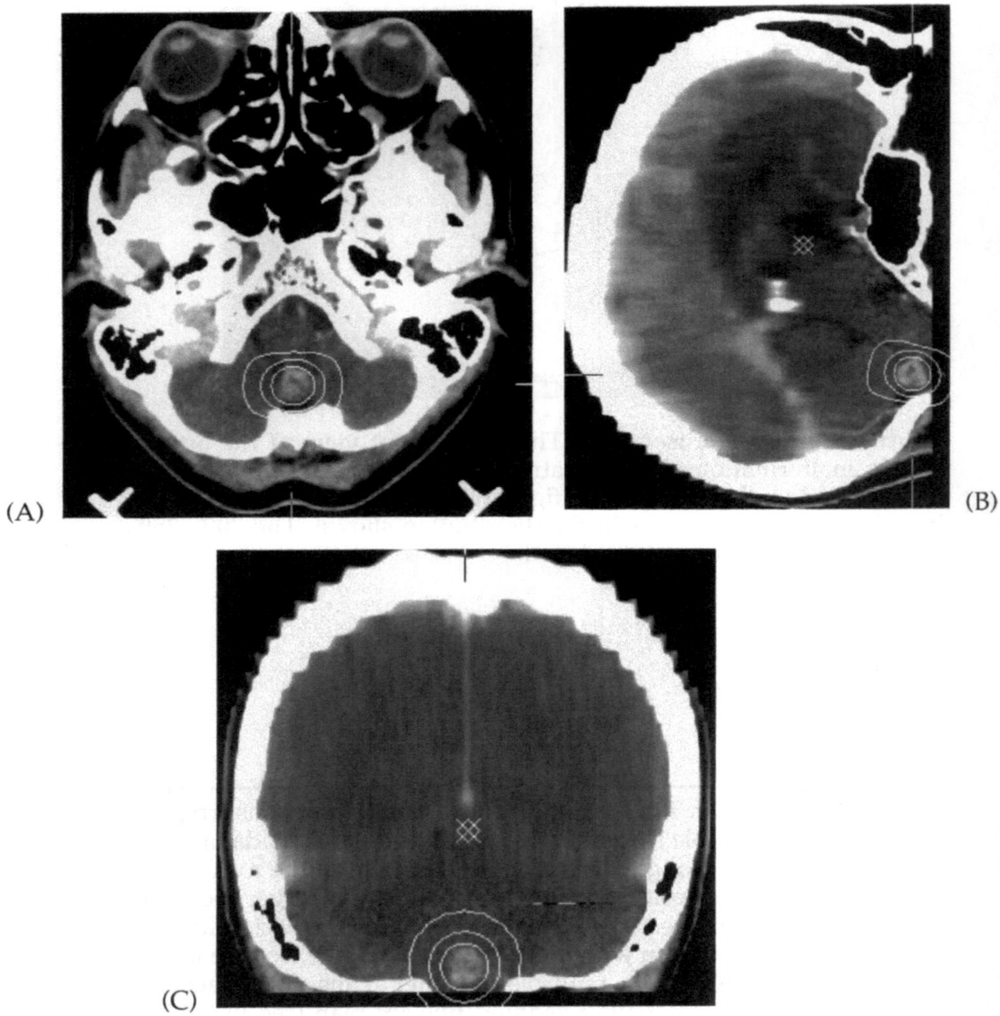

Figure 6.24. Critical neural structure anterior to the lesion. The results of the standard nine-arc plan are shown (A) in the axial plane, (B) in the sagittal plane, (C) in the coronal plane, and (D) schematically. By eliminating the anterior 20 degrees of each arc (see Fig. 6.11) we obtain the distribution shown in the axial (E), sagittal (F), and coronal (G) views. This change is shown schematically in (H). Note the elongation of lower isodose lines in the superior–inferior and medial–lateral directions with contracting of these lines in the anterior–posterior direction.

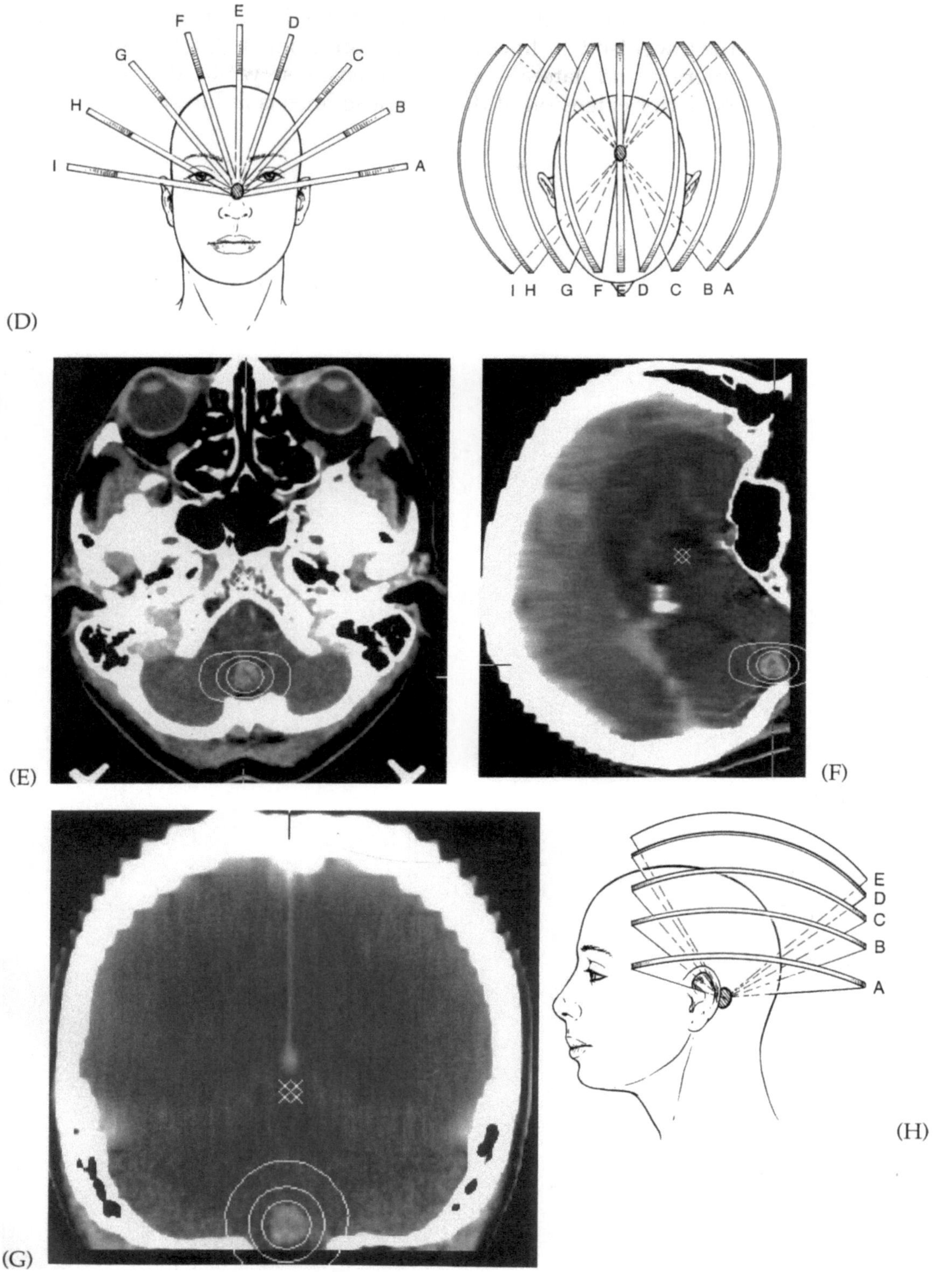

Figure 6.24. *Continued*

brain stem treated. It also illustrates the Jell-O® principle in that the lower isodose lines are now more elongated in the superior–inferior and lateral–medial dimensions, while the anterior–posterior extension of these isodose lines is reduced.

7

Dose Selection

After a treatment plan is optimized, the radiation dose is selected. In general, the dose is expressed in the unit called *gray* (Gy). One hundredth of a Gy (centigray or cGy) is equivalent to an older unit of absorbed radiation, the *rad*. The dose is usually prescribed to the isodose line (or shell, in reality) that conforms to the periphery of the target (i.e., the lesion to be treated). For example, the AVM shown in Figure 7.1 was treated with 12.5 Gy to the 80% isodose line. As seen in the figure, the 80% isodose line corresponds to the periphery of the lesion nidus. In this situation, the maximum delivered dose, or 100% of the dose (which lies near the center of the lesion), is 25% higher than the prescribed dose at the 80% isodose line, or is equal to approximately 15.6 Gy. The lower the isodose line to which the treatment dose is prescribed, the greater the difference between the prescribed treatment dose and the maximum dose; in other words, the greater the dose inhomogeneity across the target. For example, if the dose is prescribed at the 50% isodose line, the maximum (100%) dose is twice as high. All doses mentioned in this chapter are those prescribed to the periphery of the lesion.

Dose selection depends upon a detailed understanding of the radiosurgical literature; many papers provide historical dose guidelines for different radiosurgical situations. The following factors are important in dose selection: lesion volume, lesion location, preexisting neurologic deficit, proximity to radiosensitive structures, lesion pathology, and previous treatments. The prescription of dose requires a close interaction between the neurosurgeon, radiation oncologist, and radiation physicist.

We will give practical recommendations for dose selection, based on our best interpretation of the literature, as modified by our own experience with radiosurgery, in the text.

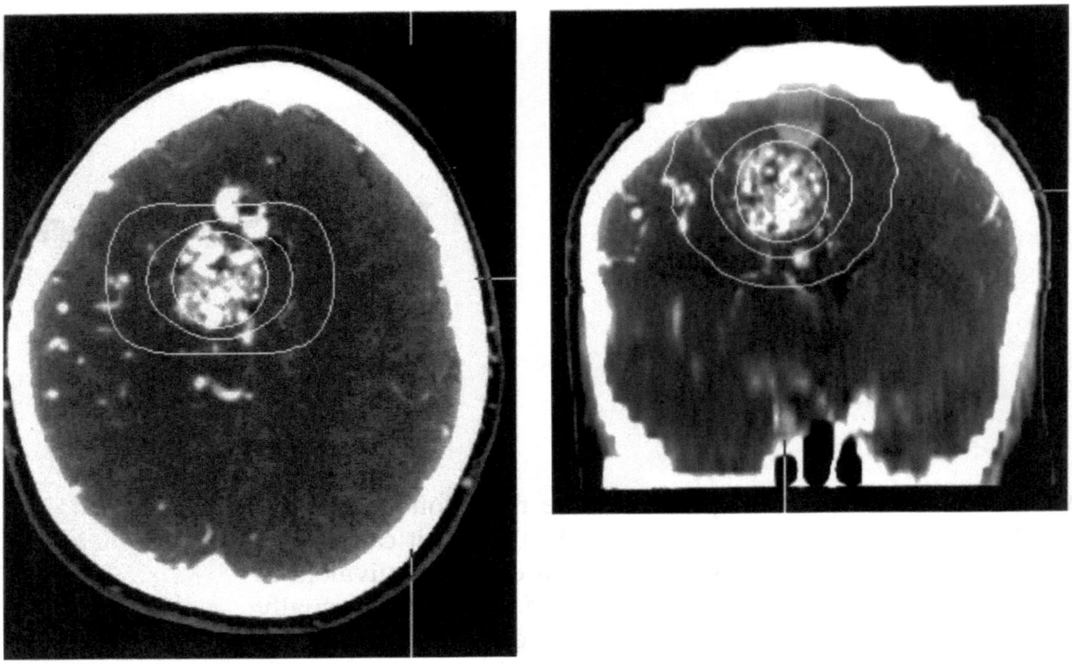

(A) (B)

Figure 7.1. An AVM treated with 12.5 Gy to the 80% isodose shell. Note that the isodose line conforms to the AVM nidus in both the axial (A) and coronal (B) views such that prescribing to the 80% line only includes nidus. The remaining lines shown are the 40% and 16% isodose lines, which will receive 6.25 Gy and 2.5 Gy, respectively. The maximum dose within the nidus will be 15.6 Gy and the minimum, which is our prescription dose, 12.5 Gy.

General Principles

1. *First, do no harm.* In general, select the lowest dose that provides the desired therapeutic effect. Selection of a higher dose may lead to more rapid response to the radiosurgical treatment, but it also increases the risk of radiation-induced complications.

2. *Dose–volume relationships.* In radiosurgery, larger lesions are usually treated with lower doses, which seems counterintuitive. However, as with conventional radiotherapy, the limiting factor in stereotactic radiosurgery is the tolerance of normal tissues to the treatment. Recall that typical radiosurgical dose gradients reduce the dose to half of the prescribed level over a distance of 2 to 3 mm. Thus, a 2-mm shell of normal brain tissue immediately adjacent to the lesion is exposed to high levels of radiation. The volume of this shell increases exponentially with the radius of the lesion treated. The dose prescribed, therefore, must be lowered as the treatment volume increases to avoid excessive complications.

A number of graphic relationships of prescribed dose-to-lesion volume or diameter have been published. The earliest was that

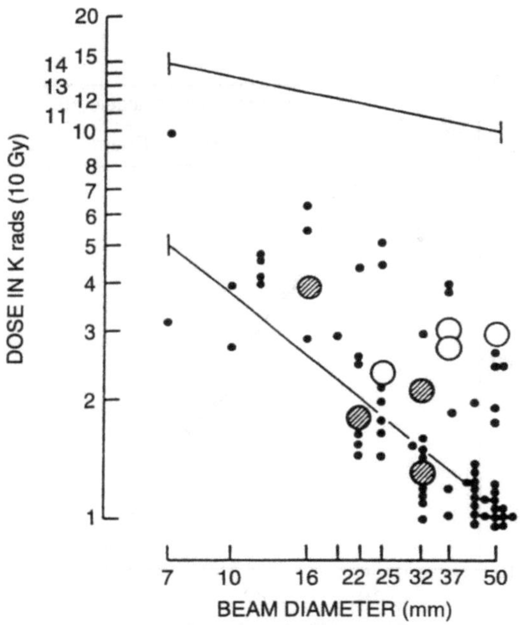

Figure 7.2. Kjellberg's curve plots the beam diameter in millimeters versus the dose in kilorads in a log–log plot. patients with complications are then noted in relationship to these factors to estimate a curve for a 1% risk of complications (lower line) versus 99% risk of complications (upper line).

of Kjellberg. Using a combination of animal and human data, he generated a logarithmic scale graph relating lesion diameter to the radiation dose expected to yield a 1% risk of complications (Fig. 7.2). This so-called 1% isorisk line on the log–log plot has been used extensively in dose selection for radiosurgery. Others, most notably Flickinger, generated dose–volume graphs based on theoretical calculations of risk, determined by basic radiobiological formulas (Fig. 7.3). Flickinger's 3% isorisk line is quite similar to Kjellberg's older 1% line. These dose–volume graphs are not concrete, but are models; the shape and location of the isoeffect line vary with the normal tissue of interest and the volume treated. In addition to these models, burgeoning quantities of actual clinical data are being produced from a number of radiosurgery centers.

Lower doses must ultimately be used to treat bigger lesions. The obvious corollary of this principle is that many lesions are too big to be safely treated with a biologically effective single dose of radiation. In general, lesions greater than 4 cm in diameter are rarely treated with radiosurgery because the dose must be reduced to a level that is unlikely to be therapeutic in a single fraction.

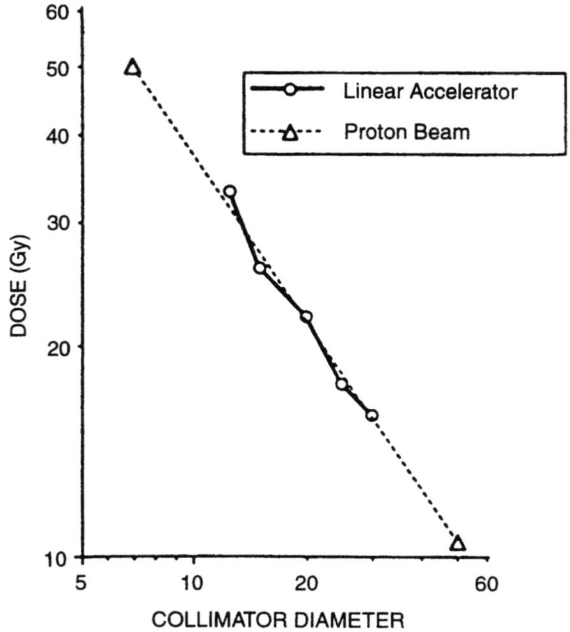

Figure 7.3. Flickinger's curve plots the collimator size in millimeters versus the dose in Gy in a log–log plot. Using radiobiologic data he estimated the 3% risk-of-complication curve (solid line), which is not significantly different from Kjellberg's initial 1% risk-of-complication curve, which is shown in the graph as a dotted line.

3. *Adjacent critical structures.* The brain is not uniformly sensitive to radiation. Optic nerves are very radiosensitive; a single-fraction dose of 7.5 Gy is currently regarded as the highest dose of radiation this structure can tolerate. Because this dose is generally not high enough to be therapeutic for tumors or AVMs, it follows that lesions abutting the optic apparatus cannot be safely and effectively treated with radiosurgery. Such lesions may be better treated with conventional, fractionated irradiation or with fractionated stereotactic radiotherapy. If the lesion is 4 to 5 mm distant from the optic apparatus, however, then an effective dose can usually be delivered to the lesion while the dose to the optic nerve remains below the 7.5 Gy limit.

The facial, trigeminal, and cochlear nerves are also relatively radiosensitive, although not as sensitive as the optic nerves (see acoustic schwannoma treatments). Motor cranial nerves that travel through the cavernous sinus (i.e., oculomotor, trochlear, and abducens) and the jugular foramen (i.e., glossopharyngeal, vagal accessory, and spinal accessory) are considered relatively radioresistant.

4. *Lesion location.* Just as in conventional neurosurgery, some regions of the brain are more risky to treat than are others. For

example, an AVM located in the right frontal lobe is generally treated with a higher dose of radiation than is a similar-sized lesion located in the motor strip, brain stem, or hypothalamus. The rationale for this different dose is that although radiation-induced brain edema or necrosis develops in a small percentage of patients, its occurrence in a "silent" brain area is likely to be asymptomatic; however, its occurrence in an "eloquent" brain area is likely to produce significant neurologic problems. Lesions in "eloquent" brain areas are consequently often treated with a slightly lower dose of radiation.

5. *Preexisting neurologic deficit.* Complications from conventional surgery or radiosurgery are more likely to develop in patients with preexisting neurologic deficits. For example, a patient with marginal seventh nerve function after previous surgery for acoustic schwannoma may have a higher risk of seventh nerve complications after radiosurgery than would a patient with normal function. Patients with preexisting neurologic deficits, therefore, may be treated with lower radiation doses.

6. *Previous treatment.* If the patient has had any prior radiation treatment, downward adjustments in radiosurgical dose may be required. Patients with malignant tumors are frequently treated with conventional radiotherapy first before undergoing a radiosurgery boost treatment. It is particularly critical that a knowledgeable radiation oncologist participate in dose selection when any prior radiation treatment has been delivered.

Guidelines for Dose Selection

Arteriovenous Malformations

The prescribed peripheral dose versus AVM nidus volume for the University of Florida series is shown in Figure 7.4. Note that larger volumes are generally treated with lower doses. We never use a peripheral dose below 10 Gy and rarely use a peripheral dose above 20 Gy. The median prescribed dose is approximately 15 Gy. The variability in dose prescribed for a given volume is caused by the other factors mentioned earlier (i.e., location, proximity of vital structures, neurologic deficit, etc.).

Acoustic Schwannomas and Meningiomas

When we began radiosurgery in 1988, the available literature recommended a minimum dose of 20 Gy to the periphery of acoustic schwannomas. Later, however, all North American radiosurgery centers confirmed that this dose is associated with an unacceptably high incidence of cranial neuropathy. The facial and trigeminal nerves are usually closely adherent to the capsule

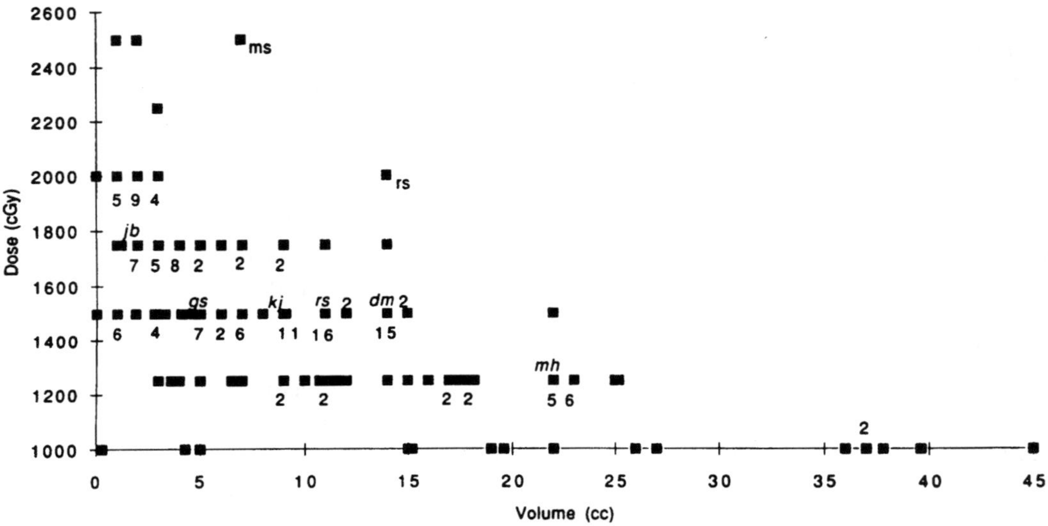

Figure 7.4. A representative linear plot of the doses used and volumes treated for our first 228 patients with AVMs showing the volume in cubic centimeters (cc) versus the dose in cGy for our initial 7 years of experience. Patients treated at each dose–volume combination are numbered. Patients with complications are noted by initials (e.g., permanent complications are shown by bold initials and temporary complications by initials in italics). Our results have agreed with both the Flickinger and Kjellberg curves with both patients above that curve treated early in our experience (ms and rs) suffering permanent complications. We initially updated these curves on a twice-yearly basis and currently update them yearly.

of acoustic schwannomas. Thus, despite the steep dose gradient associated with radiosurgery, these nerves receive essentially the same dose as does the tumor capsule. The single-fraction doses previously recommended frequently injure these nerves (at least transiently). Like most other radiosurgeons we have consequently reduced doses to the range of 10 to 15 Gy, with a much lower incidence of complications and, thus far, very high tumor control rates. We currently recommend a dose of 15 Gy for small lesions (<20-mm diameter), 12.5 Gy for medium-sized lesions (20 to 30 mm), and 10 Gy for the few lesions we treat that are larger than 30 mm in diameter. Our prescribed peripheral dose versus acoustic schwannoma treatment volume along with complications is shown in Figure 7.5. Of course, these dosage guidelines may be modified by other factors.

In general, dose recommendations for meningiomas have followed a similar history as they have for acoustic schwannomas. Doses initially reported were significantly greater than 20 Gy, and high complication rates were observed. Therefore, lesions are now generally treated to doses between 10 and 15 Gy based on the volume, location, and other modifying factors. These doses appear to produce acceptable local control and complica-

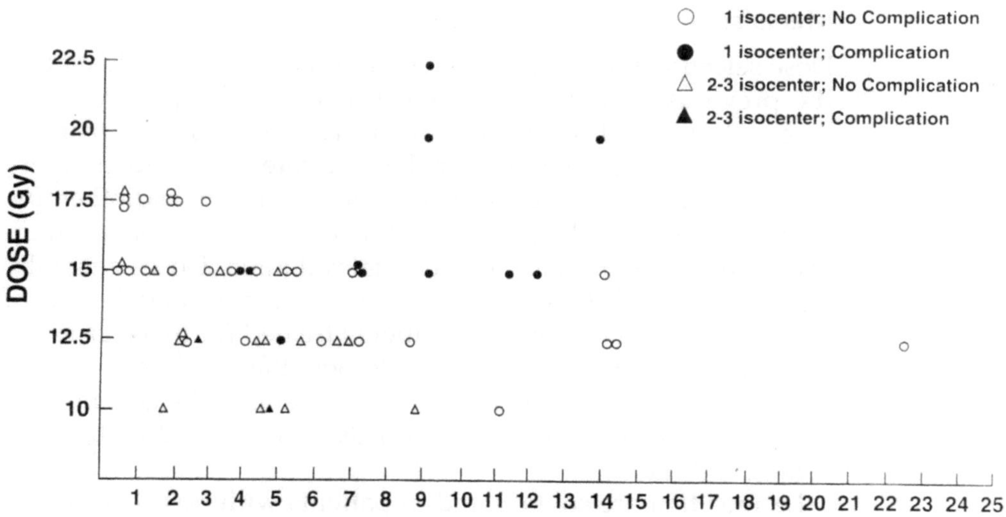

Figure 7.5. The linear dose (Gy) versus treated volume (cc) plot for acoustic schwannomas. Patients with complications are noted with solid figures. As expected, the patients treated early in our experience with high doses and large volumes have a higher complication rate.

tion rates for most small meningiomas. Lesions adjacent to radiosensitive neural structures are preferably treated with fractionated stereotactic radiotherapy (see Chapter 11).

Malignant Gliomas

Malignant gliomas, although poorly curable, display an incremental dose–response relationship. Increases in external beam dose are hampered by the large volume of normal tissue this technique mandates. Improvement in results with further escalation of radiation dose was initially reported using brachytherapy techniques that implant radioactive seeds into the target volume. More recently, stereotactic radiosurgery has been shown to be an acceptable and less invasive method for achieving a radiation boost. Because the dose–response is felt to be incremental, we attempt to maximize the radiosurgery boost dose.

Our patients generally receive accelerated hyperfractionated external beam radiotherapy to a dose of 60 Gy in 40 fractions on a twice daily schedule before radiosurgery. If the lesion is ≤2 cm, then a dose of 15 Gy is generally chosen, and if >2 cm, then a dose of 10.0 to 12.5 Gy is chosen. Modifiers such as the patient's neurologic status and the location of the lesion are also considered.

Metastases

Dose selection for metastases demands a thorough evaluation of the previous radiation treatment because many patients have received external beam treatment as well as the planned radiosurgery. Some patients are being treated with radiosurgery as an adjuvant boost to their primary brain radiotherapy, whereas others are being treated for recurrence at some significant interval from their initial cranial radiation. Doses generally vary in terms of total dose, dose per fraction, and length of treatment time. Furthermore, patients treated for metastases are quite heterogenous in terms of histology, the number of lesions, size of lesion or lesions, and functional status. In general, our doses vary between 10.0 and 17.5 Gy based on all of the preceding factors.

Lesions that are solitary and in patients with no other known disease are generally treated with 15.0 to 17.5 Gy unless the lesion is greater than 3.0 cm or near a critical structure. Multiple lesions at disparate locations can usually be treated as indepen-

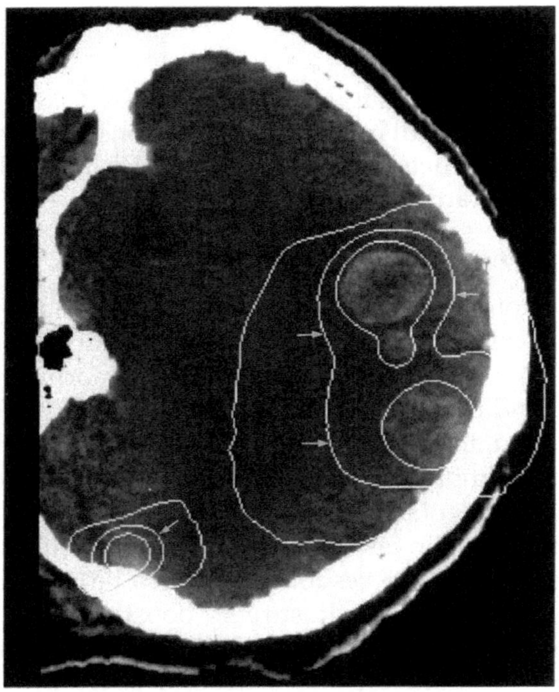

Figure 7.6. A patient with four metastases treated with radiosurgery. Note that several lesions in close proximity create a large volume of normal brain treated to 50% or more of the prescription dose (arrows). Such large volumes mandate a lower prescribed dose to the lesions despite the individual sizes of the metastases.

dent lesions. However, lesions are often so close together, that the contribution of dose from each to the normal brain results in a larger-than-expected volume of normal brain receiving a high dose. It is therefore important when choosing a dose that as many of the lesions as possible be accounted for in a common plane and that the composite dose to the normal brain be considered (Fig. 7.6).

The bottom line. Dose selection is a complicated process that requires close communication among the neurosurgeon, radiation oncologist, and radiation physicist. A careful consideration of the lesion volume, location, histology, previous radiation dose, and neurologic status is important.

8

Radiation Delivery

General Characteristics of Linear Accelerators

Treatment Beam Generation and Collimation

Linear accelerators (linacs) are the most common source of therapeutic radiation. Radiation is produced by rapid acceleration of electrons through a microwave wave guide followed by their bombardment of a heavy metal target. These high-energy electrons are stopped in the target and lose most of their energy as heat, but they also lose a small portion of their energy as X-rays (see Fig. 6.1). The X-rays produced have a range of energies with a maximum equal to the energy of the accelerated electrons. The average X-ray beam energy produced is approximately one third that of the maximum energy. A linac is named after the maximum energy of an X-ray it produces. The most common machines used for stereotactic radiosurgery have maximum energies between 4 million electron volts (MV) and 15 MV. At the University of Florida we use a 6 MV linac for stereotactic radiosurgery. A more complete description of radiation beam production is available in several references in the bibliography.

Once the radiation beam is produced, the next step in treatment delivery is to refine the beam so that it is suitable for patient treatment. More specifically, a radiotherapy beam must be homogeneous over defined areas that comprise the treatment field. To achieve this, the beam is both collimated to define edges or field sizes and filtered to make the beam flat and even (i.e., more uniform) over the treatment field area (see Fig. 6.1). This collimating and filtering of the beam is also more fully discussed in bibliography references.

For radiosurgery, treatment fields that are used are generally only 0.5 cm to 4.0 cm in size, whereas the fields are generally significantly larger in standard radiotherapy. Because of the decreased field sizes and the desired increase in accuracy of beam delivery, the radiosurgery treatment field is further defined by

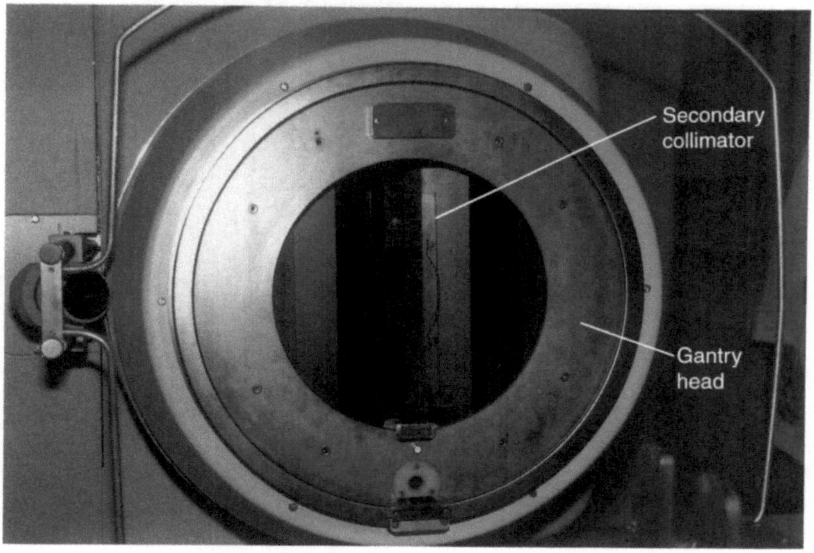

(A)

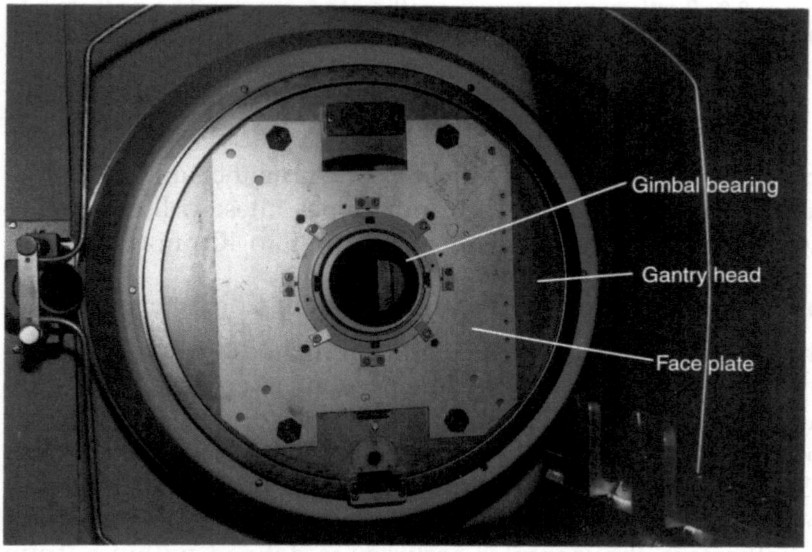

(B)

Figure 8.1. (A) The secondary or adjustable linac collimators are shown as they appear looking into the gantry head (see also Fig. 6.1). Note that these collimators are two pairs of adjustable jaws (in the x and y direction) that allow formation of variable square or rectangular fields. (B) The face plate for the tertiary radiosurgery collimators attaches to the gantry head. This face plate contains a gimbal bearing that corrects for gantry sag (i.e., improves accuracy), and the collimators shown in Figure 8.2 fit within the aperture of this gimbal bearing on the face plate.

an extra (i.e., tertiary) collimator (Figs. 6.1, 8.1, and 8.2). These tertiary collimators are generally circular, provide a steeper dose gradient at the beam edge than do standard collimated beams,

Figure 8.2. Tertiary radiosurgery collimators ranging in size from 5 to 40 mm.

and allow improved centering of the treatment beam. In order to treat a variety of lesion sizes and shapes, these collimators are available in 2- to 5-mm increments from 5 to 40 mm (Fig. 8.2).

Treatment Machine and Treatment Room Characteristics

The relationship of the refined and collimated X-ray beam to the given target within the treatment room is critical. Linacs are mounted on a rotating gantry such that the beam has a center of rotation approximately 1.4 to 1.6 m above the finished floor (Fig. 8.3). The intersection of the collimator axis on the axis of gantry rotation is defined as the isocenter of the linac. For standard radiotherapy, the isocenter accuracy is defined within a 2 mm sphere.

In addition to gantry rotation, the treatment table shown in Figure 8.4 rotates about the isocenter. It is capable of moving vertically, laterally, and longitudinally to position the patient such that the target center (in a patient) is located at the isocenter of the linac. This table also has an accuracy of rotation that is defined within a 2-mm sphere. Hence, the combined mechanical inaccuracy of a standard radiation therapy linac is 4 mm at a maximum.

Some linacs are more accurate than are others. Although the rotating gantry must have an isocenter defined within 2 mm for

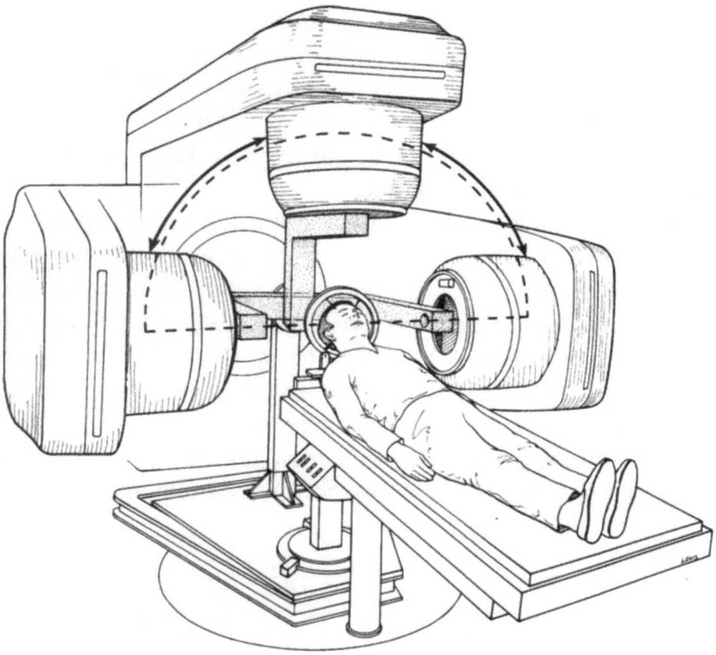

Figure 8.3. The linac gantry rotates in the indicated manner around the isocenter.

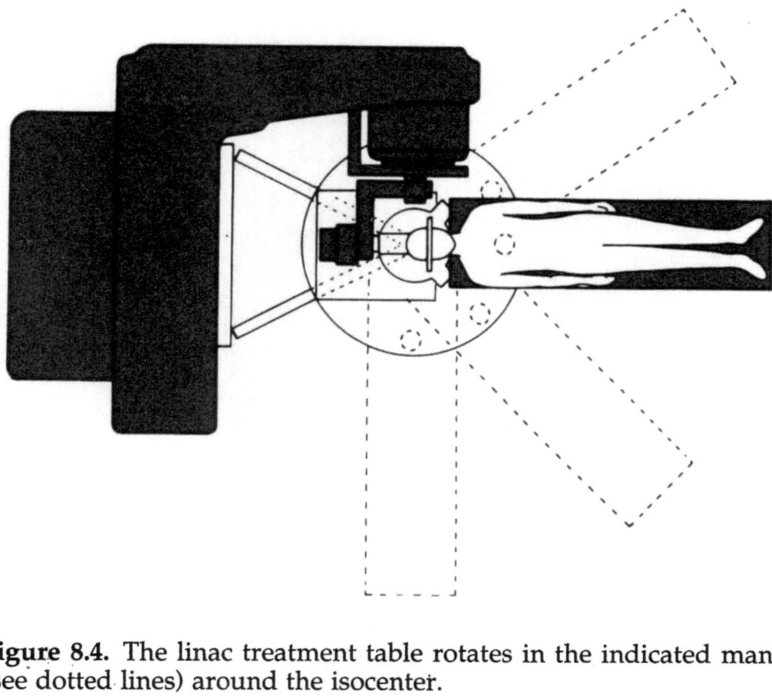

Figure 8.4. The linac treatment table rotates in the indicated manner (see dotted lines) around the isocenter.

standard therapy, some machines may be substantially better or worse. The addition of tertiary collimators may make gantry sag worse than expected. Because of stereotactic radiosurgery, one company (Varian, Palo Alto, CA) has designed a lighter weight gantry apparatus that minimizes gantry sag. Table accuracy may be worsened by patient loading.

Because stereotactic radiosurgery depends on optimized accuracy, an improved system was designed at the University of Florida for stereotactic radiosurgery. This system improved accurate focusing of the beam on the isocenter by adding a set of bearings to the stereotactic collimator system that accounts for imperfections in the gantry rotation. In addition, a set of bearings attached to the patient and target area was achieved in the floor subsystem that bypasses the inaccuracies of the table rotation (Fig. 8.5). Together, this system achieves mechanical accuracy

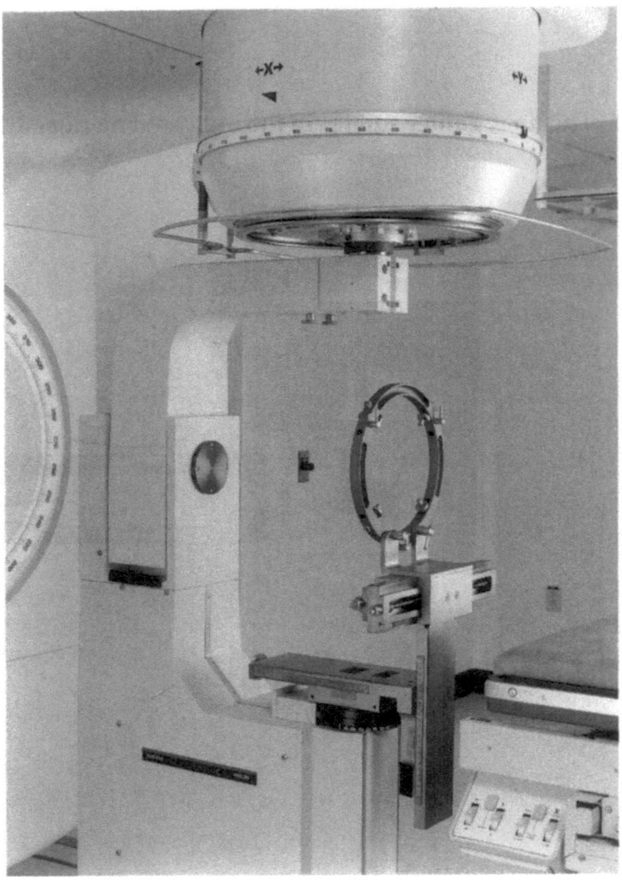

Figure 8.5. The University of Florida radiosurgery subsystem corrects for gantry sag and table rotation inaccuracies so that isocenter for the purpose of radiosurgery may be defined within 0.2 mm ± 0.1 mm.

within 0.2 mm ± 0.1 mm for defining the treatment isocenter of beam delivery.

It is important to thoroughly test the mechanical accuracy of treatment delivery before treatment with a radiosurgery system. The modified test system proposed by Lutz and Winston is used at the University of Florida, and its use during daily treatment delivery is described in the next section.

A Typical Stereotactic Radiosurgery Treatment Delivery

This section outlines our procedures for stereotactic radiosurgery treatment delivery. These procedures will vary among specific linac radiosurgery systems, but the basic ingredients of an initial hardware modification for conversion from external beam radiotherapy to radiosurgery, position verification of the treatment target, and a well-defined sequence for patient treatment delivery are common to all systems.

Radiosurgery Treatment Delivery Setup

Before treatment, the standard radiotherapy treatment linac must be converted to a radiosurgery treatment device by attachment of the radiosurgery isocentric subsystem. This requires 8 to 12 minutes by our treatment team, which generally consisting of five to six people. Our procedure is:

1. Remove the standard radiotherapy treatment table top and replace it with the shorter cushioned stereotactic radiosurgery table top (Fig. 8.6A).
2. Rotate the linac's collimator to 0 degrees and set the field size to 8.0 × 8.0 cm.
3. Attach the face plate with gimbal bearing onto the head of the gantry (Fig. 8.6B).
4. Remove the floor cups and wheel the radiosurgery subsystem into position over the floor base rings (Fig. 8.6C).
5. Lower the wheels, setting the subsystem into position, and then remove the wheels (Fig. 8.6D).
6. Attach the microswitch interlock cable (Fig. 8.6E). This interlock prevents gantry rotation into the radiosurgery subsystem base, thereby preventing damage to the system as well as patient injury.
7. Attach the ram collar lock to the table support ram (Fig. 8.6F). This collar is to prevent failure of the ram table support during treatment, which would be dangerous for a patient attached to the stereotactic subsystem. The ram collar lock is tightened to prevent changes in the table height after the patient is in proper position for treatment.

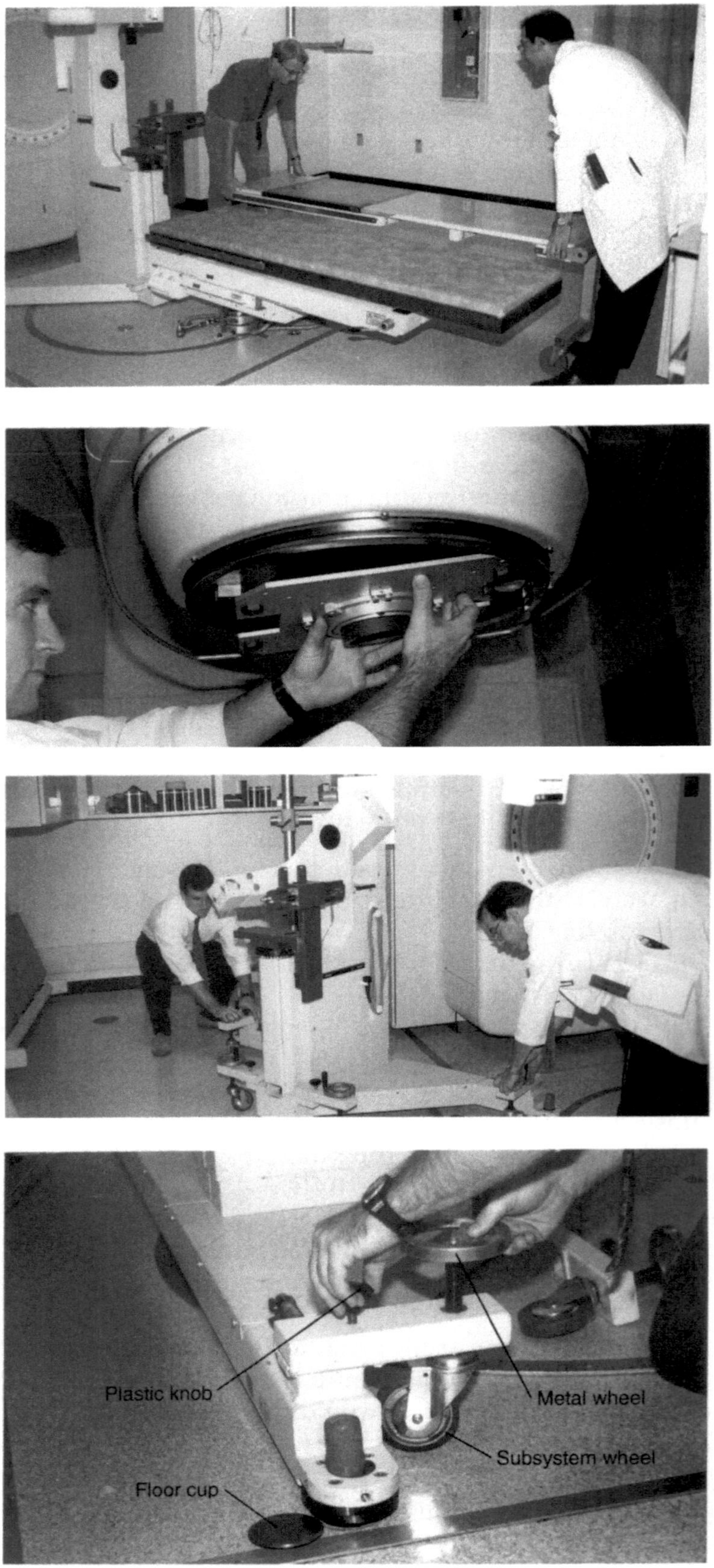

(A) (B) (C) (D)

Figure 8.6 (A) The treatment table for daily radiation treatments is removed by two members of the treatment team and replaced by the cushioned couch. This couch is shorter to allow the subsystem to rest at the head of the radiosurgery treatment table. (B) The face plate with the gimbal bearing is attached to the head of the linac gantry with four screws. (C) The radiosurgery subsystem is being wheeled into position in front of the linac gantry. (D) After removing the floor cups, the subsystem is lowered onto the floor rings that hold it in proper position in relationship to the gantry. This is achieved by turning the metal wheel shown. Finally, these wheels are removed by turning the plastic knob on the wheel attachment. (E) Microswitch attachment. (F) Attachment of collars to the table support ram. (G) Raising the collimator arm beneath the gimbal bearing. (H) Inserting a collimator so that it lies flush with the collimator holder of the collimator arm and within the aperture of the gimbal bearing. (I) Closing the collimator door and screwing it shut.

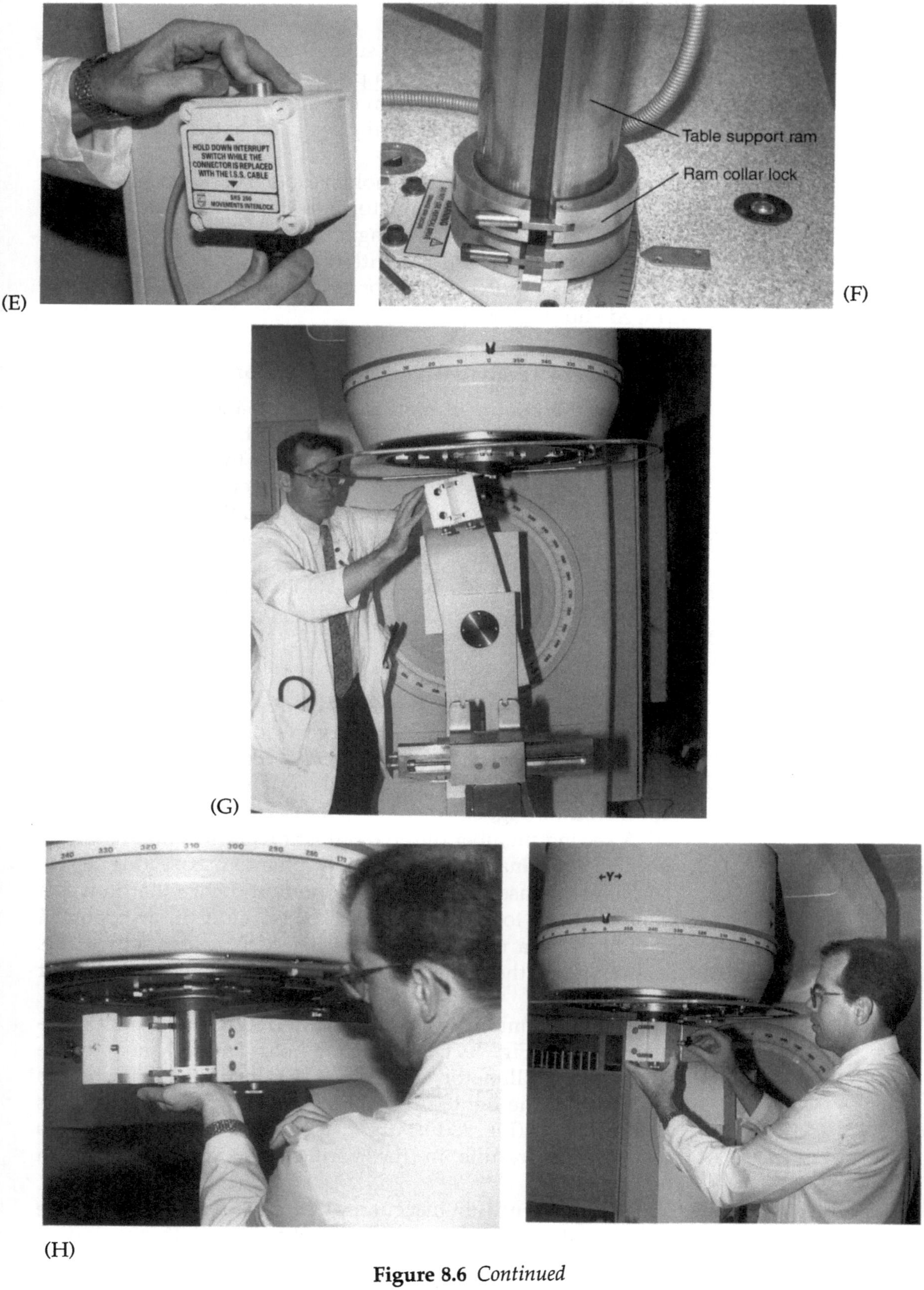

(H)

Figure 8.6 *Continued*

8. Remove the pin from the subsystem collimator arm while maintaining the collimator arm position, and then slowly allow the arm to rise so that the collimator arm is beneath the collimator opening in the gimbal bearing attached to the linac's gantry head (Fig. 8.6G).

9. Unscrew the collimator holder door screw and open the holder door. Insert the collimator into the gimbal-bearing collimator hole, and allow it to rest against the collimator holder wall and shelf so that it is flush within the collimator holder (Fig. 8.6H). The door of the collimator holder may then be closed and screwed shut (Fig. 8.6I).

Radiosurgery Treatment Target Verification

Now that the treatment plan is generated and the room is assembled, the system is ready for treatment. To deliver a treatment, the target lesion must be placed at the isocenter of the machine. Before delivering treatment, a verification system is used to assure that the treatment isocenter defined in the plan is indeed that which is set on the subsystem, and that the subsystem relationship to isocenter is stable and hence will deliver the treatment to that properly defined target accurately. The verification system used is a modification of the Lutz-Winston system.

1. One member of the team sets the treatment isocenter coordinates on the stereotactic radiosurgery subsystem. The coordinates are established on vernier scales on the subsystem that enables the setting of the AP, lateral, and vertical coordinates to within 0.1 mm (Fig. 8.7A).

2. A second member of the treatment team sets the treatment isocenter coordinates into the BRW phantom base (Fig. 8.7B). The phantom base is a critically important device that moves a phantom point to the treatment plan isocenter in an identical relationship to a mock BRW ring (Fig. 8.7B). After setting this phantom point, the phantom target is approximated to that point (Fig. 8.7C).

3. The phantom target is then transferred to the patient treatment system (Fig. 8.7D). The film holder is attached to the radiosurgery collimator arm (Fig. 8.7E). Inspection first occurs by looking at the lighted treatment field in relationship to this phantom target (Fig. 8.7F). After inspection, a series of four films is taken that results in the verification film illustrated in Figure 8.8.

In our experience, an inaccuracy of greater than 0.2 mm on the treatment film is obvious to a trained observer. Methods to quantitate the error for a set of films are described in the bibliography, but they are beyond the scope of this chapter. We suggest that any initial experience include film measurement until the operators are confident than an unacceptable error will be detected by inspection.

Radiosurgery Treatment Delivery

1. If the treatment collimator is not 24 mm in size, then the collimator is changed to meet the treatment plan specification.

2. The patient is brought into the treatment room and sits on the treatment couch. The patient is asked to tilt the head forward, and the alignment bolts are loosely inserted into the undersurface of the BRW ring (Fig. 8.9).

3. The patient then is raised on the treatment couch to approximately the correct position and is then asked to lie down with the head extending over the edge of the treatment couch (Fig. 8.10A) for attachment to the radiosurgery subsystem. The patient is gently lowered into position and firmly attached to the subsystem (Fig. 8.10B).

4. At this point, the patient is in position for treatment, and the final interlocks are activated. The table ram that controls height is disabled by its switch, and the ram collar lock is tightened as a second method of assuring that the table height is fixed throughout treatment (Fig. 8.11A). The mechanical locks for the lateral and axial table movement are also tightened so that they will not move during the treatment (Fig. 8.11B).

5. The treatment procedure is again explained to the patient, and the first table position and gantry rotation are set according to the prescription. All positions are verified by two independent persons on the treatment team, and the radiation is then delivered to each specified arc in the sequence. A typical treatment plan prescription is shown in Figure 8.12. Note that next to each step there is a blank space that is checked by the independent observers before delivery of the radiation for each arc used in the treatment plans.

6. If a collimator is changed in the middle of a treatment plan, then we recommend that the gantry be rotated away from the patient so that the risk of accident during the change is minimized.

7. If a multiple isocenter plan is used, then the subsequent isocenters are set on the isocentric subsystem without releasing the patient. In order to move the patient more than a few millimeters, however, the table position and sometimes the table height may need to be changed. The interlocks are therefore removed and the patient is adjusted to maintain the body in comfortable position in relationship to the head. After achieving

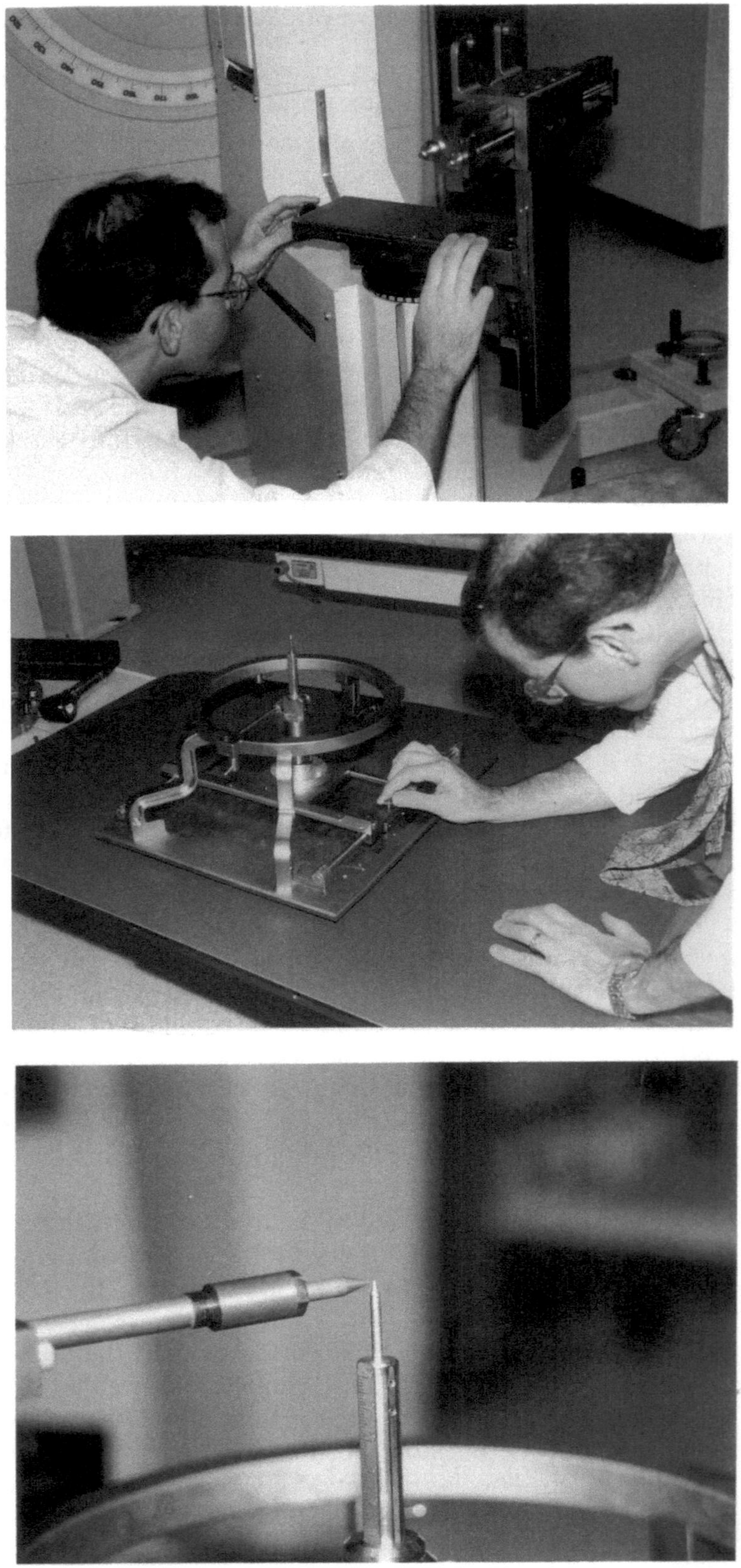

(A)

(B)

(C)

Figure 8.7. (A) Radiosurgery team member setting the target coordinate on the stereotactic radiosurgery subsystem. (B) The phantom base has a mock-BRW ring. A phantom point is set to the target coordinate using the same vernier scales that are present on the radiosurgery subsystem. The team member is setting the phantom point. (C) The phantom target is approximated to the phantom base as shown. The point used in this approximation is then replaced with a solid target, which is clearly visible on verification films. (D) The phantom target is then transferred to the radiosurgery subsystem, which is set to the identical target coordinate. If the targets are set correctly and the system accuracy is not degraded, then the subsequent verification procedure confirms the accuracy of the system and the correct coordinate setting for the patient. (E) Attachment of the film holder. (F) Light verification. The light field of the linac is initially activated and should visually confirm that the target is in the approximate center of the field. Target rotation is also done during the light verification and is a good pretest of accurate target placement before film verification.

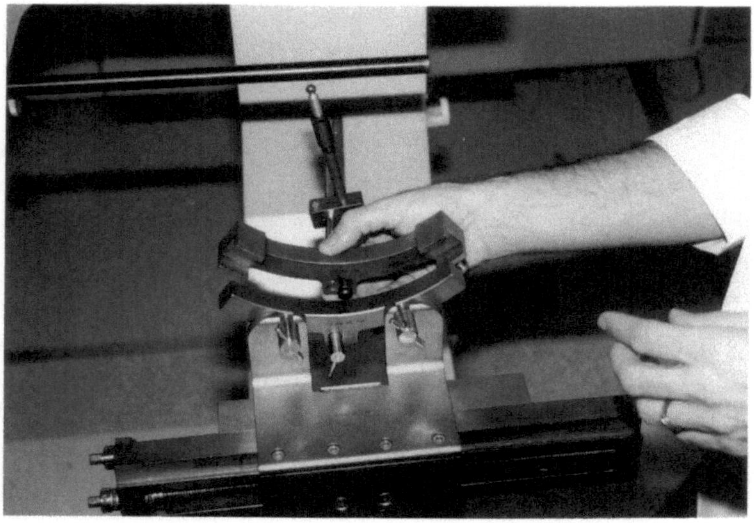

(D)

(E)

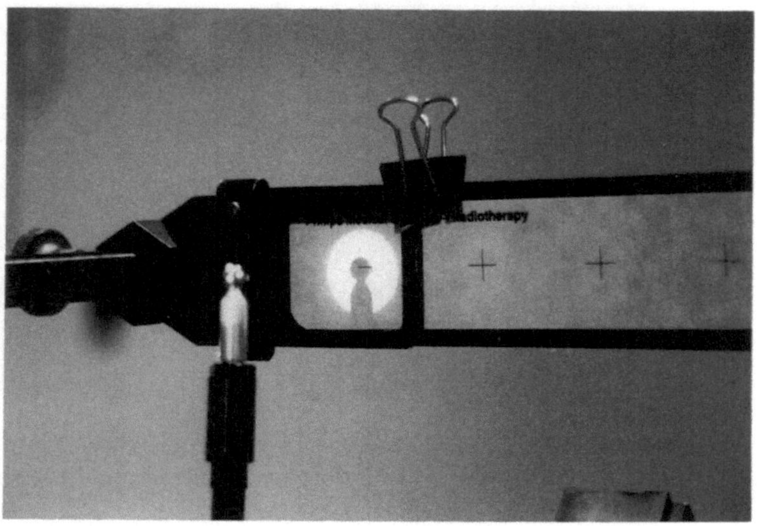

(F)

Figure 8.7. *Continued*

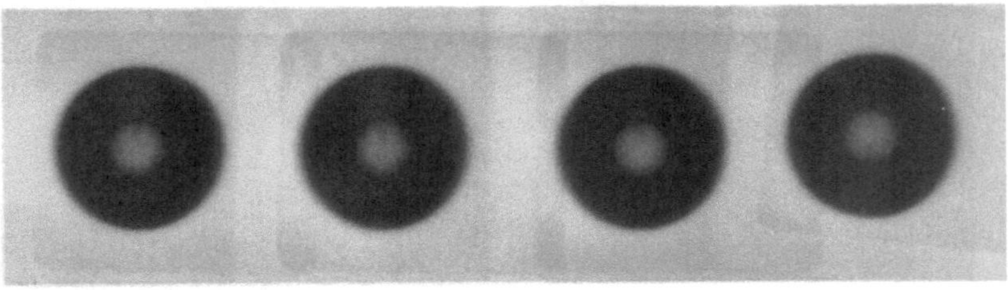

Figure 8.8. Film verification. A series of four films is taken of the target with a 24-mm collimator at different gantry angles and table rotations to confirm that the target is placed appropriately in the AP, lateral, and axial directions.

the correct position for the new isocenter, the interlocks are all reinstituted. After repositioning the patient, the new treatment isocenter is confirmed by independent readings of two other members of the treatment team. If all of the readings agree, then the treatment of the new isocenter proceeds as the first.

8. After treatment of all isocenters is complete, the patient is released from the stereotactic radiosurgery subsystem and allowed to sit up on the treatment couch. The table interlocks are released and the patient is lowered into a comfortable position for the neurosurgeon to remove the BRW ring. The pins are removed and the ring is released. Any slight bleeding that occurs with the removal is managed with local pressure, and the patient is then permitted to leave.

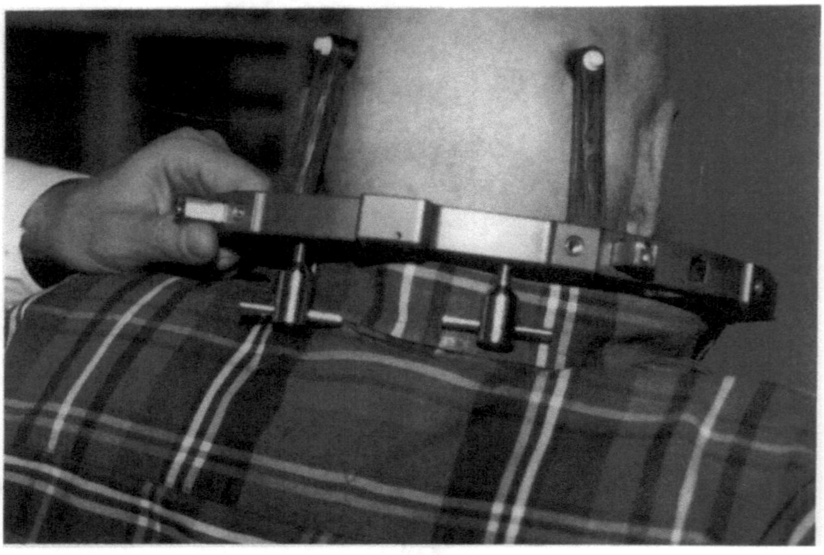

Figure 8.9. The alignment bolts again placed in the undersurface of the patient's BRW ring to prepare for attachment to the radiosurgery subsystem.

The bottom line. Radiosurgery treatment delivery requires a carefully designed verification system and checklist. The preceding example applies to the University of Florida system. Table-mounted systems and different floor-mounted systems have different requirements for verification, but all require diligent quality assurance during treatment. A checklist system is essential.

(A)

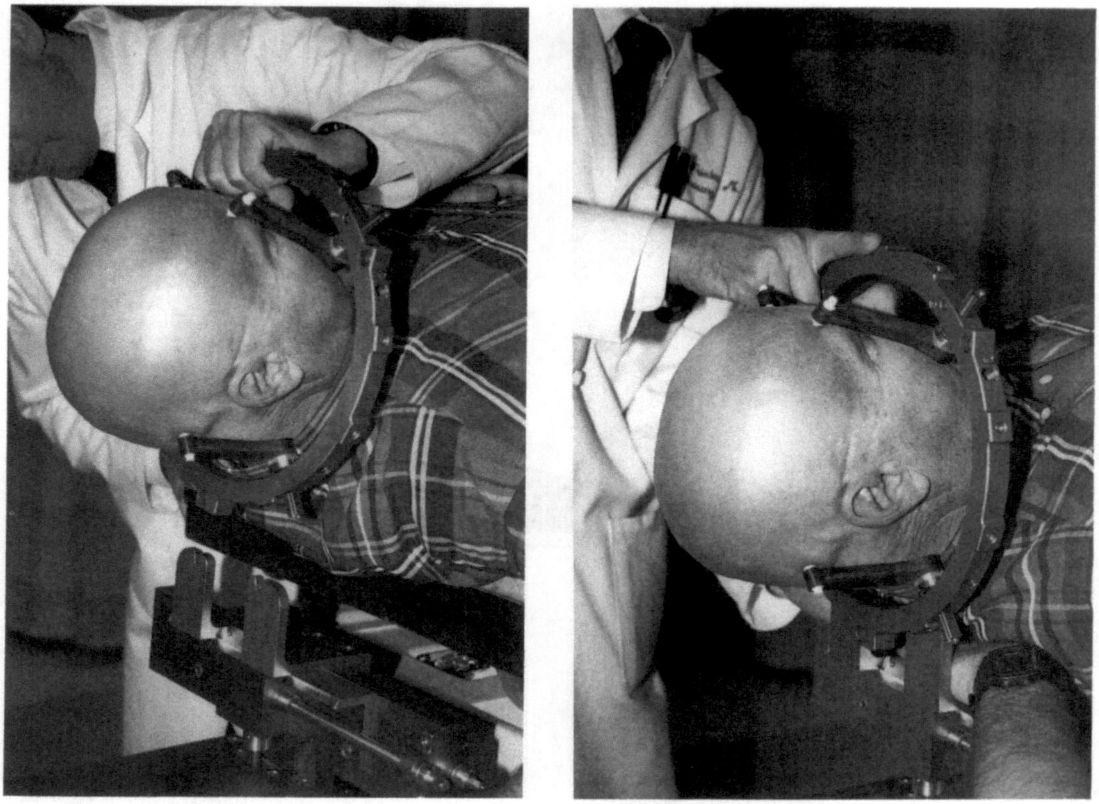

(B)

Figure 8.10. (A) The patient lies down, head extending over the edge of the treatment couch. (B) The patient is lowered so the alignment bolts mate with the radiosurgery subsystem; the bolts are then tightened to complete patient attachment.

(A)

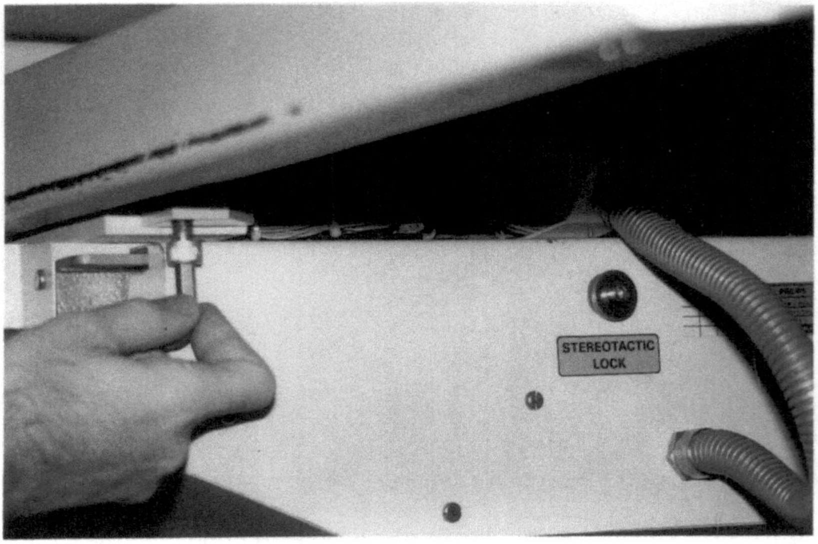

(B)

Figure 8.11. (A) The ram collar is tightened after the table ram support switch is inactivated. (B) Table locks are tightened to prevent axial and lateral movement.

```
Date : Tue Aug 13 08:26:27 EDT 1996

Patient Name :
Prescribed dose : 1500.0

Percent Line    : 80.0

Peak value of dose matrix      (max) =   903.270
Treatment Plan Rept Factor (repwt) =      2.076

----------------------------------------------------------------

          Setup and Validation Procedure

____ Collimator rotation set to 0 degrees.

____ Field size set to 8 x 8 cm.

____ Collision avoidance connected.

____ Patient wrist band ID checked prior to treatment

____ Ram isolation turned off.

____ Ram table top locked in x and y directions.

     Take test film shots:
            Gantry     Stand
     _____   270         0
     _____   230       315
     _____   230       270
     _____   330       270
     _____   330       315
     _____   90          0
     _____   130        45
     _____   30         45

____ Visual approval of test film

____ Treatment plan compared to FDA approved system.
----------------------------------------------------------------

____   1.  Set coordinates to A-P =  -39.9
                              Lat =   -9.2
                              Axl =  -60.4

____   2.  Install 12.0 mm collimator.

____   3.  Set patient table to   10 degrees.

____   4.  Treat : Start angle = 130
                   End angle   = 30
                   Mon units   = 308

----------------------------------------------------------------

____   5.  Set patient table to   30 degrees.

____   6.  Treat : Start angle = 130
                   End angle   = 30
                   Mon units   = 316

----------------------------------------------------------------
```

Figure 8.12. (A) A typical radiosurgery treatment checklist has two components: an initial verification checklist and a checklist of specific instructions for treatment. Each step is performed by one member of the radiosurgery team and independently verified by two other members of the team.

9

Patient Follow-Up

Radiosurgery is unlike most conventional neurosurgical procedures because results of treatment are unknown when the treatment is completed. The neurosurgeon is accustomed to examining the patient in the recovery room and knowing very rapidly whether the immediate outcome of a procedure is good or bad. Radiosurgery has few, if any, acute complications. The only acute complication seen in our patients has been a rare increase in seizure activity in the first 48 hours after radiosurgery in patients that previously exhibited seizure activity. Patients go home the day of treatment exactly as they were before treatment. The true efficacy of the treatment and incidence of complications can only be known after months or years of careful follow-up.

Physicians delivering radiosurgery treatment must therefore be committed to rigorous follow-up procedures. In general, we see our patients or request information on them regularly, but at intervals that vary according to the disease treated. Our recommended follow-up schedule for commonly treated lesions is shown in Table 9.1.

Site-Specific Recommendations

AVM patients have follow-up along with MRI/MRA scans at 12-month intervals until the AVM resolves. An angiogram is then performed at 3 years to confirm thrombosis. Angiograms are necessary because MRI/MRA results are relatively unreliable as the nidus shrinks to less than 1 cm in size. If no nidus is present on MRI/MRA, a repeat angiogram is scheduled at 3 years. If it shows a small residual, retreatment is usually performed that day. A patient with significant residual nidus, however, may be retreated without angiogram or be treated with microsurgery/embolization.

Acoustic neuroma and meningioma patients are seen at 12-month intervals with an MRI every 12 months as well, for a

Table 9.1. Radiosurgery patient follow-up recommendations.

Type of lesion	Follow-up*	Imaging
AVM (benign)	every 12 mo for 3 years	MRI/MRA every 12 mo until lesion is resolved or at at 3 yr. After resolution or 3 yr an angiogram confirms ablation or is used for retreatment planning
Acoustic schwannoma (benign)	every 12 mo	12 mo for a minimum of 5 yr (then as clinically indicated)
Meningioma (benign)	every 12 mo	12 mo for a minimum of 5 yr (then as clinically indicated)
Pineocytoma (benign)	every 12 mo	12 mo for a minimum of 5 yr (then as clinically indicated)
Pituitary adenoma (benign)	every 12 mo	12 mo for a minimum of 5 yr (then as clinically indicated)
Gliomas (malignant)	every 3 mo	MRI at 3 mo intervals for 2 years and then every 6 months
Metastases (malignant)	every 3 mo	MRI at 3 mo and every 6 mo thereafter for a minimum of 2 yr, then every 12 mo

*All patients should have follow-up indefinitely at yearly intervals except as noted.

minimum of 5 years of radiographic follow-up. At that point, follow-up is maintained on a yearly basis with imaging dictated by clinical symptoms.

It is necessary to obtain more frequent CT or MRI follow-up on patients with malignant lesions, such as metastases or gliomas. For metastases, a scan to assess the response of the lesion is made 3 months after the radiosurgery, and then every 6 months thereafter for a minimum of 3 years. Malignant gliomas are generally scanned every 3 months for the first 2 years, and then every 6 months. Scan frequency is often dictated by clinical symptoms, although yearly scanning beyond 2 years is reasonable. A low threshold for repeat imaging is maintained in patients with malignant lesions because tumor growth may rapidly lead to patient deterioration and further treatment might be considered for a small recurrence.

Follow-up for other, less common, lesions, such as pituitary or pineal tumors, should mirror the clinical and radiographic follow-up after radiotherapy. Yearly follow-up in some fashion should be pursued for all patients after radiosurgery because late disease recurrence and complications are known to occur many years after standard radiotherapy.

At the same time that evidence of therapeutic efficacy is sought with follow-up imaging, clinical information must be obtained to verify that complications have not occurred. The preferred assessment method is to examine the patient personally. If this is not possible, then information is obtained from the patient's local physician or directly from the patient. It is important to detect complications early so that appropriate treatment measures can be expeditiously instituted. For example, steroids may produce dramatic improvement in patients with radiation-induced brain edema. As another example, a patient with a transient seventh nerve palsy needs instruction pertaining to eye care to prevent corneal ulceration while the nerve is recovering.

Case Examples

The following are detailed vignettes of some of our complicated cases to illustrate the importance of careful follow-up for radiosurgery patients.

Patient 1

A 55-year-old woman in whom a grand mal seizure disorder developed was discovered to have a right mesial frontal AVM (Fig. 9.1A). She was treated with 25 Gy to the 80% isodose line through a 24-mm collimator. At 1 year, an angiogram showed complete thrombosis. Seizures and a left hemiparesis developed 1 month later. T2-weighted MRI scans revealed an area of probable radiation necrosis in the exact area treated with radiosurgery, surrounded by considerable edema (Fig. 9.1B). She was treated with steroids, which produced a prompt and dramatic clinical improvement. After several months of therapy, the steroid dosage was tapered. Follow-up MRI revealed minimal abnormality in the treatment area (Fig. 9.1C). Our current knowledge of appropriate doses for AVMs of a given size now leads us to prescribe a significantly lower dose for a similar-sized AVM (see Fig. 7.4).

Patient 2

At presentation, a 30-year-old woman had a seizure disorder, was found to have an AVM in the motor strip area on the left side, and was referred for radiosurgery (Fig. 9.2A). The AVM was treated with 15 Gy to the 80% isodose line through a 26-mm collimator. At 1 year, angiography revealed substantial, but incomplete, thrombosis of the lesion. The patient had headaches 14 months after treatment. MRI revealed an area of gadolinium enhancement at the treatment site, with marked surrounding

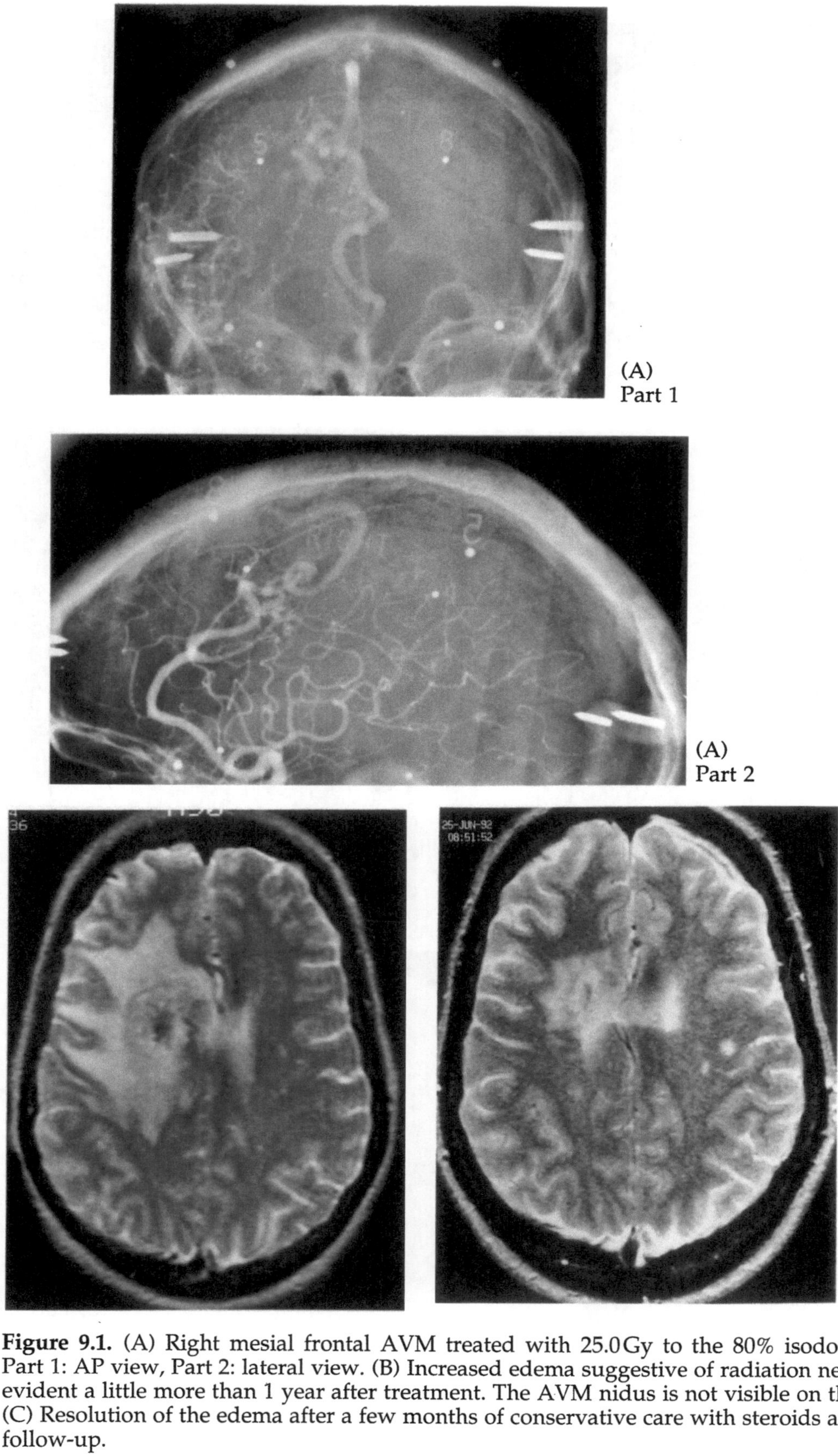

Figure 9.1. (A) Right mesial frontal AVM treated with 25.0 Gy to the 80% isodose shell. Part 1: AP view, Part 2: lateral view. (B) Increased edema suggestive of radiation necrosis is evident a little more than 1 year after treatment. The AVM nidus is not visible on this scan. (C) Resolution of the edema after a few months of conservative care with steroids and close follow-up.

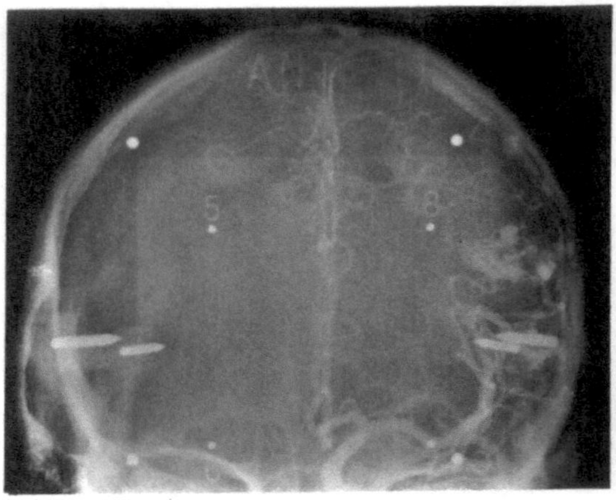

(A) Part 1

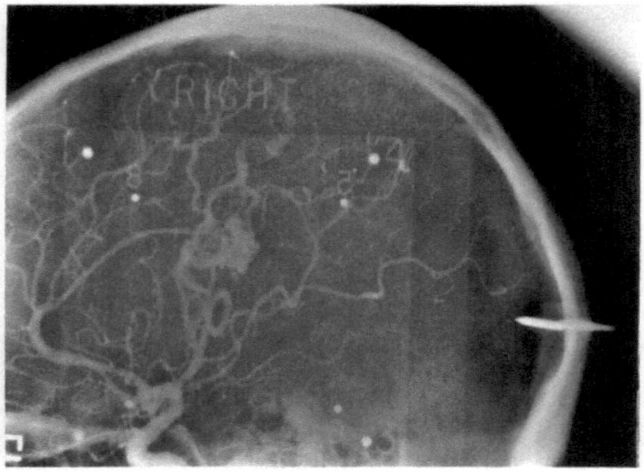

(A) Part 2

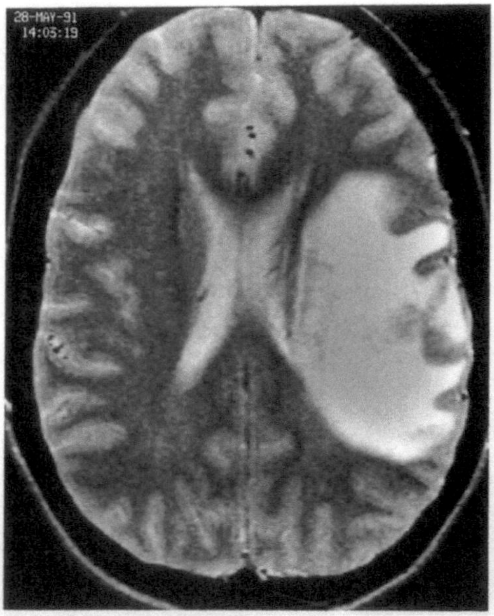

(B)

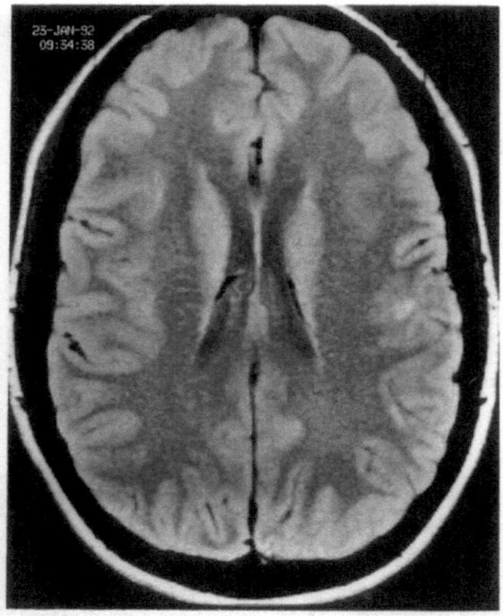

(C)

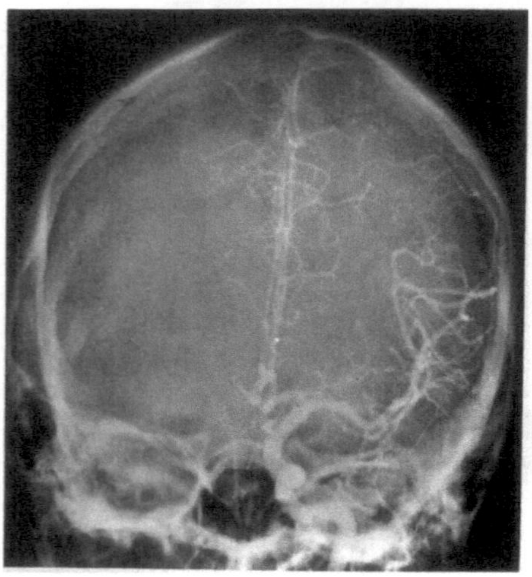

(D) Part 1

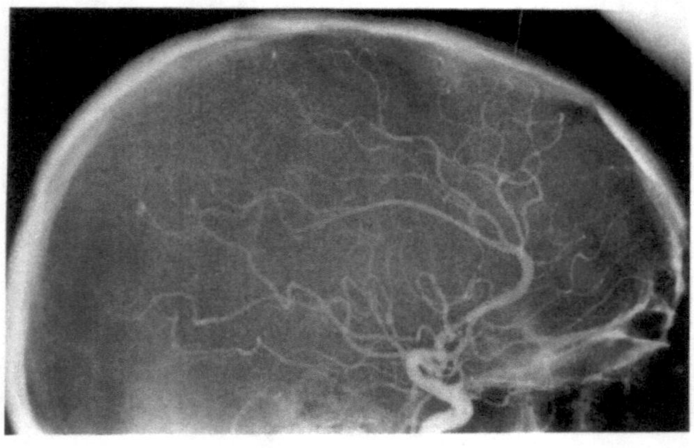

(D) Part 2

edema (Fig. 9.2B). Her headaches responded to dexamethasone treatment, and she continued taking decreasing doses of steroids for 3 months. Follow-up MRI revealed complete resolution of the edema (Fig. 9.2C), and subsequent angiography revealed complete thrombosis of the AVM (Fig. 9.2D).

Patient 3

A 61-year-old man underwent clipping of a carotid aneurysm and partial resection of an AVM in the region of the right motor strip (Fig. 9.3A). He was referred for radiosurgical treatment of the residual nidus. He received 15 Gy to the 80% isodose line through a 22-mm collimator. The patient had an intracerebral hemorrhage with significant hemiparesis 3 months after radiosurgery (Fig. 9.3B). After prolonged rehabilitation, he recovered to his pretreatment level. An angiogram 1 year after treatment revealed substantial, but incomplete, thrombosis (Fig. 9.3C). Follow-up angiography performed 4 years after his initial treatment revealed complete thrombosis of the AVM (Fig. 9.3D).

When examined parenthetically, the greatest drawback of radiosurgery is that it does not appear to provide a protective effect against hemorrhage until the AVM is completely obliterated. Because this typically takes 1 to 3 years, the patient remains at risk for a substantial period of time. In contrast, the risk of hemorrhage can immediately be eliminated with microsurgical AVM excision.

Patient 4

A 69-year-old man with a long history of hearing loss had a CT scan revealing a 30-mm-diameter acoustic schwannoma (Fig. 9.4A). He was treated with 20 Gy to the margin of his lesion. Approximately 1 year after treatment, a complete facial nerve palsy developed. The paralysis improved but had not completely resolved 2 years after treatment. Follow-up CT revealed loss of internal enhancement and tumor shrinkage (Fig. 9.4B). This dose of radiation, originally recommended in the literature, is now widely recognized to be too high. A dose of 12.5 Gy would currently be chosen. Use of lower doses led to a dramatic reduction in the incidence of facial and trigeminal nerve complications after radiosurgery. In addition, cranial neuropathy in

←

Figure 9.2. (A) Left motor strip AVM. Part 1: AP view; Part 2: lateral view. (B) Increased edema 14 months after treatment with 15.0 Gy to the 80% isodose shell. (C) Resolution of edema after three months of steroids. (D) Angiogram confirming sucessful outcome. Part 1: AP view; Part 2: lateral view.

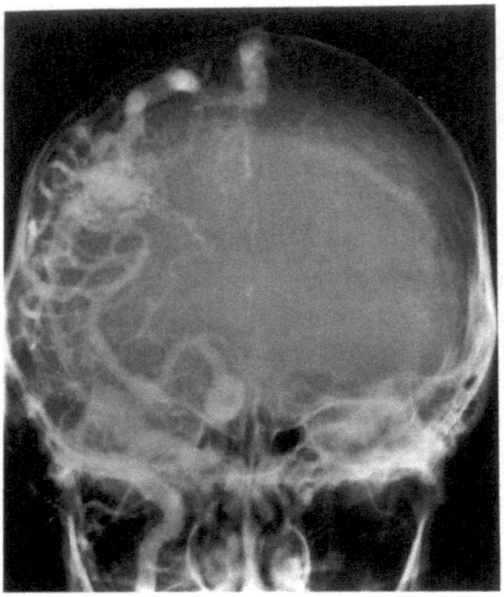

(A) Part 1

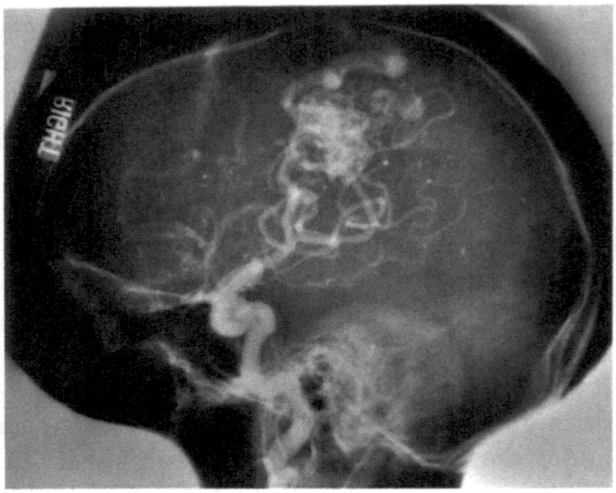

(A) Part 2

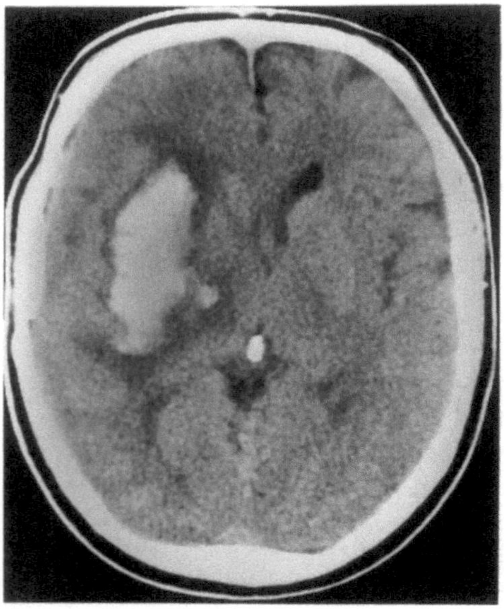

(B)

Figure 9.3. (A) Right motor strip AVM. Part 1: AP view; Part 2: lateral view. (B) Scan showing intracerbral hemorrhage 3 months after radiosurgery. (C) Imaging 1 year after treatment showing partial resolution of the AVM nidus. Part 1: AP view; Part 2: lateral view. (D) Repeat angiogram confirming AVM ablation 4 years after treatment. Part 1: AP view; Part 2: lateral view.

patients with tumors smaller than 24 mm in diameter has rarely been seen in our series.

The following general principles are helpful for treating radiation-induced complications.

1. The most important method of managing complications is to prevent complications from ever developing, through precise

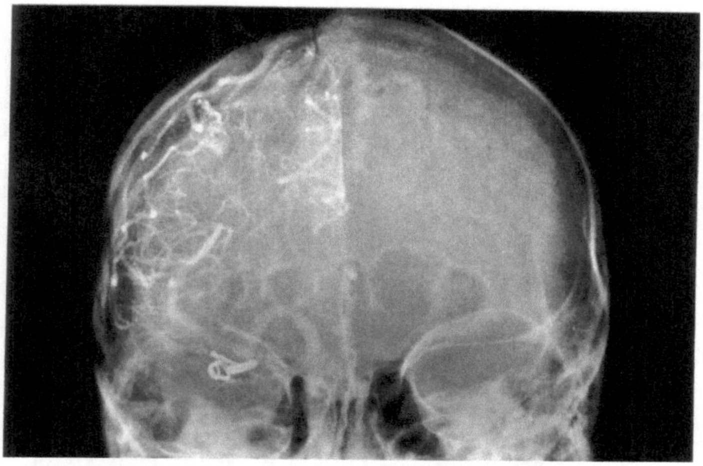

(C)
Part 1

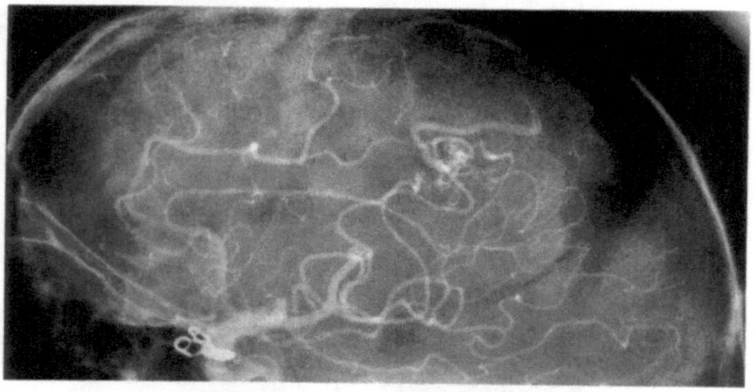

(C)
Part 2

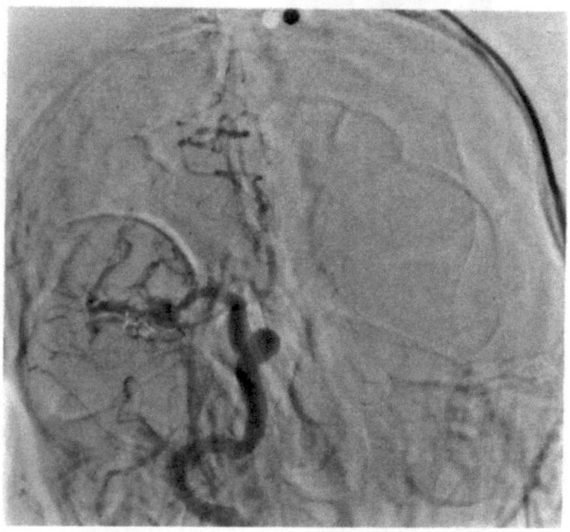

(D)
Part 1

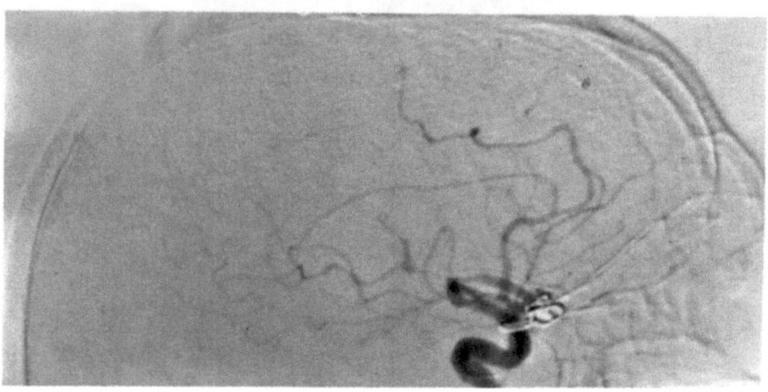

(D)
Part 2

Figure 9.3. *Continued*

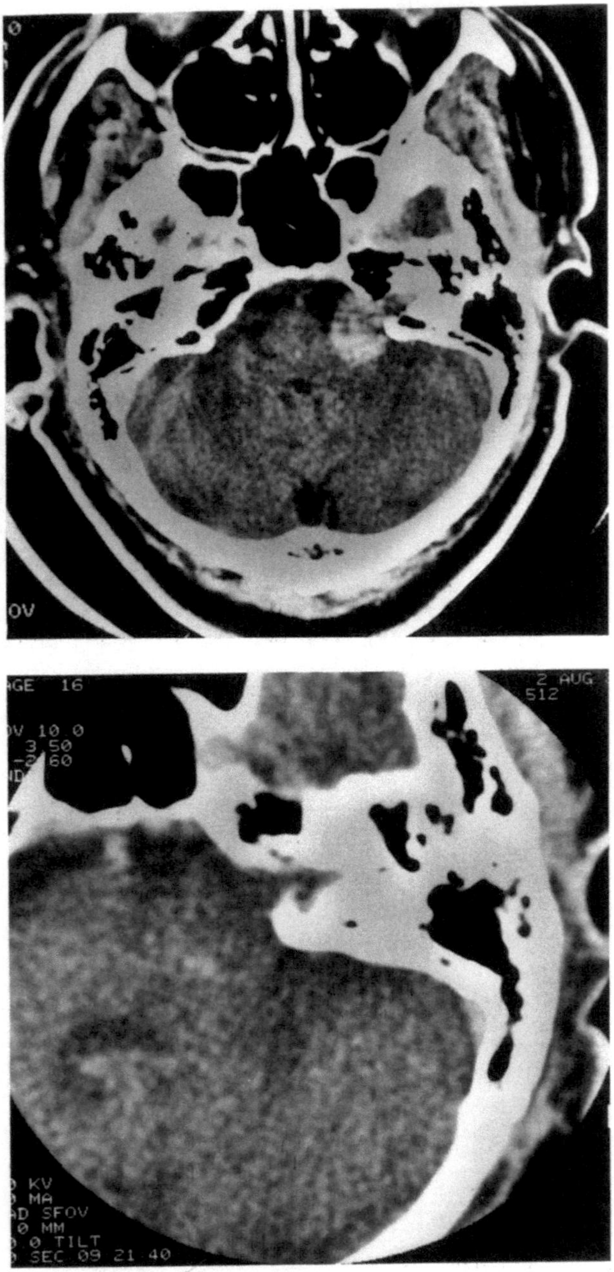

(A)

(B)

Figure 9.4. (A) Acoustic schwannoma treated to a dose of 20.0 Gy. (B) Follow-up scan at 1 year showing decrease in size of the lesion along with central hypolucency consistent with tumor necrosis.

dosimetry planning, accurate dose delivery, and selection of the lowest effective radiation dose. *It is critically important to avoid including normal tissue in the radiosurgical treatment field.*

2. If principle No. 1 is diligently pursued, then radiation-induced complications are almost always transient. They typically occur after a latent period of 6 to 14 months and run a course to total or near-total resolution in 3 to 6 months. It is important to reassure patients that the problem is usually transient.

3. Steroids are valuable in relieving the symptoms of radiation-induced brain edema. We typically use dexamethasone, starting with 4 mg four times a day and then steadily taper the dosage to the lowest dose that is consistent with normal neurologic function. Frequent neurologic examinations and radiographic procedures are often helpful in guiding therapy.

4. To improve the results of radiosurgery, all complications must be scrupulously documented. This requires a commitment to careful follow-up of all treated patients indefinitely.

10

University of Florida Results

Arteriovenous Malformations

Between May 18, 1988, and May 21, 1996, 644 patients were treated with radiosurgery at the University of Florida. The types of lesions treated are shown in Table 10.1. Of these patients, 264 had AVMs. There were 131 men and 133 women in this group. The mean age was 39 years (range, 7 to 70 years). Presenting symptoms included hemorrhage (93 patients), seizure (102 patients), headache or incidental symptoms (61 patients), and progressive neurologic deficit (8 patients). The location of the lesions is given in Table 10.2. Spetzler-Martin classification is shown in Table 10.3. Twenty-five patients had undergone prior surgical attempts at AVM excision, and 21 patients had undergone at least one embolization procedure. Most patients referred for radiosurgery were first screened by an expert cerebrovascular surgeon, and radiosurgery was undertaken only if the patient was a poor candidate for conventional microsurgery.

The mean radiation dose to the periphery of the lesion was 15.0 Gy (range, 7.5 to 25 Gy). Dose–diameter (or dose–volume) guidelines previously described, as well as lesion location and clinical variables, were used to select the dose (see Chapter 7). In general, the larger the lesion, the smaller the dose of radiation (see Fig. 7.4). The specified dose was almost always delivered to the 80% isodose shell (range, 70 to 90%). There were 198 patients treated with one isocenter, 35 patients with two isocenters, 22 patients with three isocenters, 5 patients with four isocenters, 2 patients with five isocenters, and 2 patients with six isocenters.

Mean lesion volume was 9.6 cc (range, 0.5 to 45.3 cc), and median lesion volume was 7.2 cc. In an effort to provide data comparable to other publications in the radiosurgical literature, the following size categories were used in this analysis: A (<1 cc), B (1 to 4 cc), C (4 to 10 cc), D (>10 cc). Treatment volume was determined in all cases by performing a computerized dose–volume

Table 10.1. Lesions treated with radiosurgery at the University of Florida.

	No. Pts.	%
Arteriovenous malformations	264	41
Acoustic schwannoma	94	15
Meningioma	54	8
Gliomas	50	8
Metastases	107	17
Other*	75	12

*Including pituitary adenomas, hemangioblastoma, nonacoustic schwannoma, nasopharyngeal carcinoma, pineal tumors.

histogram of the treatment isodose shell, which was constructed to conform to the AVM nidus.

Mean follow-up duration for the entire AVM group was 42 months (range, 1 to 96 months). Follow-up often consisted of clinical examination and MRI every 6 months after treatment, although we have currently reduced our follow-up to every 12 months. If possible, follow-up was performed at the University of Florida; otherwise, MRI and examination results were obtained from the patient or the patient's local physician. Clinical information was available for 252 of 264 patients.

All patients were initially asked to undergo angiography at yearly intervals, regardless of the MRI findings. After the first 50 patients were treated, we decided to defer angiography until MRI strongly suggested complete thrombosis. Furthermore, if complete thrombosis was not identified 3 years after radiosurgery, then repeat radiosurgery or microsurgery was undertaken in an effort to obliterate any remaining nidus.

Table 10.2. Location of arteriovenous malformations.*

Location	No. Pts.	%
Frontal	60	23
Temporal	25	9
Parietal	70	27
Occipital	30	11
Internal capsule/basal ganglia	13	5
Thalamus	18	7
Brainstem	12	5
Cerebellum	8	3
Corpus callosum	8	3
Cerebellopontine angle	1	>1
Centrum	15	6
Choroidal	3	1
Other	1	>1

*Based on 264 patients.

Table 10.3. Spetzler-Martin classification.

	No. Pts.*	%
I	16	6
II	94	36
III	110	42
IV	42	16

*Spetzler-Martin grade was not noted in two patients.

Results

Outcome categories are discussed in this section, along with detailed results within each category. Results are summarized by category and AVM volume in Table 10.4.

1. Angiographic cure. In general, conventional angiography was performed when MRI or MRA suggested complete thrombosis, or if the patient had evidence of a persistent nidus 36 months posttreatment (e.g., as part of the retreatment procedure). Timing varied somewhat, depending on patient compliance and the logistics of long-distance radiographic scheduling.

Angiographic cure requires that no nidus or shunting remains on the study, as interpreted by a diagnostic radiologist and the treating neurosurgeon (Fig. 10.1). A total of 69 patients had angiographic cures out of 101 angiograms performed (69%). These patients reached a definitive successful end point for radiosurgery. The following angiographic cure rates were seen in the various size categories: (A) 33%; (B) 88%; (C) 85%; (D) 44%.

It is interesting that two patients with angiograms showing small remaining nidi at 2 years had complete occlusion on follow-up angiograms at 3 years.

Table 10.4. Results of AVM treatment by size category (A–D).

	A	B	C	D
Angiographic cure	1	29	22	17
Angiographic failure	2	4	4	22
Retreated	1	3	3	18
Deceased	—	1	1	5
MRI cure	2	2	6	6
MRI failure	—	—	2	9
Refused follow-up	—	2	3	5
Lost to follow-up	1	3	4	4

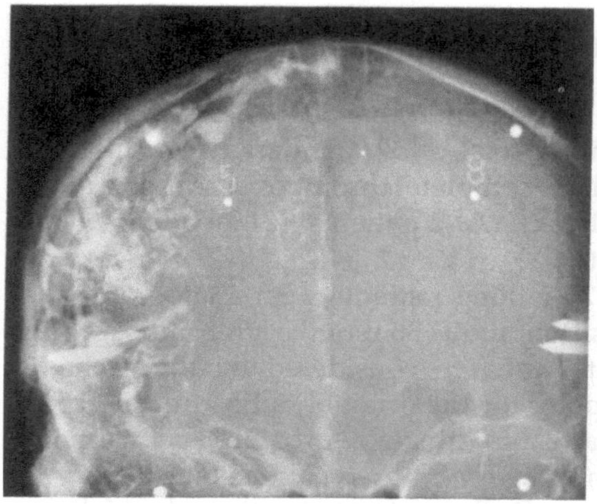

(A) Part 1

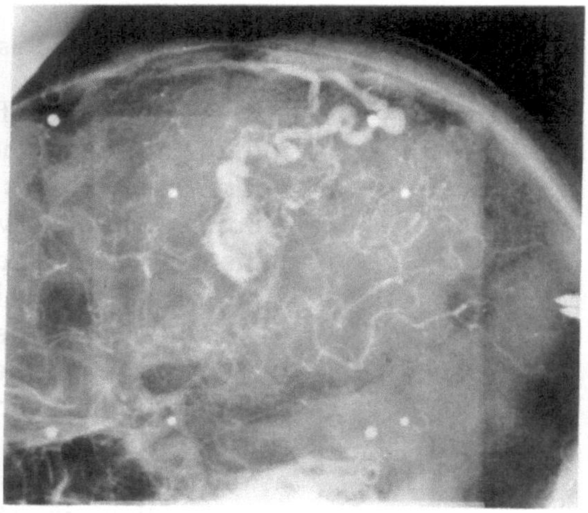

(A) Part 2

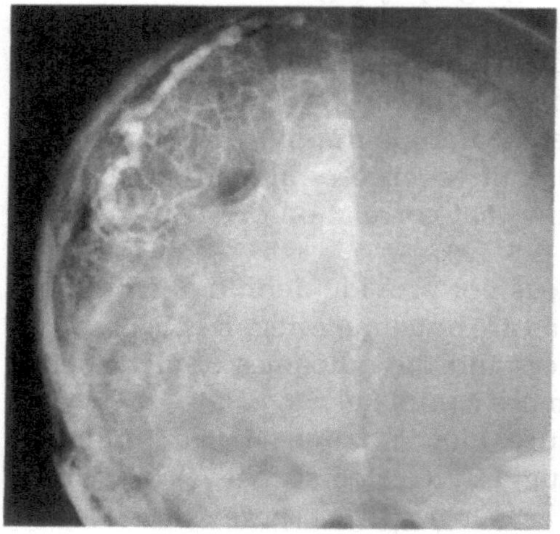

(B) Part 1

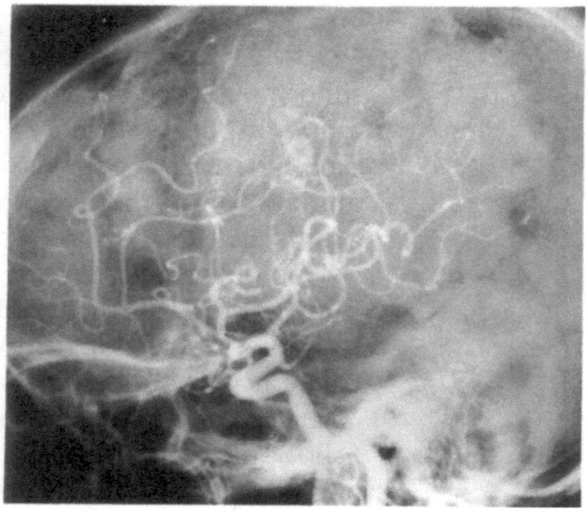

(B) Part 2

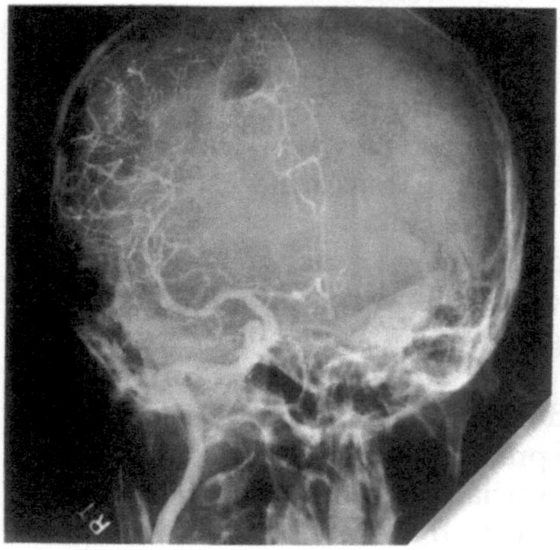

(C) Part 1

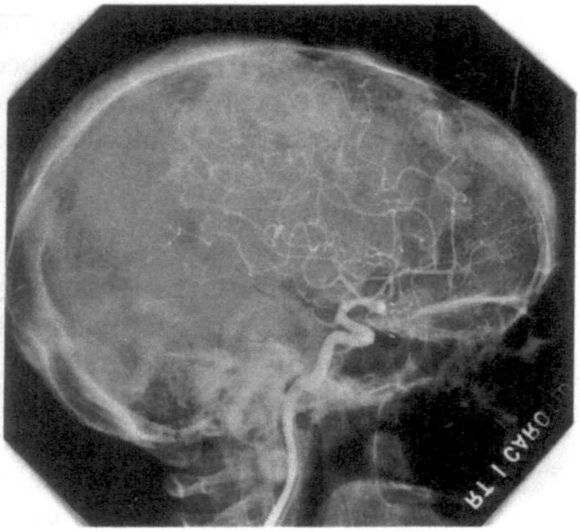

(C) Part 2

Figure 10.1. (A) Treatment-planning angiogram. Part 1: AP view; Part 2: lateral view. (B) Follow-up angiogram at 1 year. Part 1: AP view; Part 2: lateral view. (C) Angiogram at 2 years showing angiographic ablation of the arteriovenous malformation. Part 1: AP view; Part 2: lateral view.

2. Angiographic failure, greater than 24 and less than 36 months after treatment. A total of 32 angiograms in this time period showed less than complete thrombosis. Angiographic failures correlated with size: (A) 2 patients; (B) 4 patients; (C) 4 patients; (D) 22 patients.

3. Retreatment. All 25 willing patients with angiographic or MRI evidence of a persistent nidus 36 months after radiosurgery underwent retreatment. They were considered to have reached a definitive failure end point for the original radiosurgical procedure (Fig. 10.2). One patient in category A required retreatment, as did three patients in category B, three patients in category C, and 18 patients in category D. Two retreated patients had follow-up long enough to have angiography again, and both were cured.

4. MRI suggestive of cure. Sixteen patients currently under follow-up had MRI evidence of AVM thrombosis and either had angiography pending or refused angiography. Although these patients may reach the definitive successful end point (i.e., angiographic cure), we know that correlation of MRI to angiography is less than perfect.

This outcome category does not provide definitive end point information. It is included, as are outcome categories 5, 6, and 7, to allow the reader to understand the outcomes of all patients treated with radiosurgery more completely.

5. MRI suggestive of failure. Eleven patients currently under follow-up had MRI scans suggestive of persistent AVM (two in size category C and nine in category D). These studies were performed 24 to 30 months posttreatment. All studies suggested substantial but incomplete AVM thrombosis.

These patients were not considered to have reached a definitive failure end point because two patients experienced AVM thrombosis between 2 and 3 years posttreatment. When and if they reach 36 months posttreatment, they will undergo angiography and will be assigned to cure, retreatment, or follow-up categories.

6. Ten patients refused follow-up.

7. Twelve patients were lost to follow-up. As of the time of last contact, no positive or negative results of radiosurgery were identified.

8. Seven patients died during the follow-up period. In all cases, information as to the precise cause of death was obtained from the family or local physician. Five of the patients died of intercurrent disease, unrelated to the AVM or to radiosurgery.

Two patients reportedly died suddenly, with symptoms consistent with an intracerebral hemorrhage.

Outcome End-Point Summary

Definitive outcome end points included angiographic cure (category 1), retreatment (category 3), and fatal hemorrhage (category 8). Patients currently in category 2 (angiographic failure, <36 months), category 4 (MRI suggestive of cure), and category 5 (MRI suggestive of failure) may, judged by previous experience (see the discussion), eventually move into definitive success or failure end-point categories. Patients in category 6 (refused follow-up) and category 7 (lost to follow-up) are indeterminate.

Definitive outcome by size category is summarized in Table 10.5. A successful end point has thus far been attained in 50% of size category A patients, 91% of size category B patients, 88% of size category C patients, and 46% of size category D patients.

Complications

Acute Morbidity

Eight patients experienced seizures within 48 hours of radiosurgery. All but one had a seizure disorder before treatment. Anticonvulsant levels are routinely optimized in the high normal range before radiosurgical therapy. Prophylactic anticonvulsants are not administered to AVM patients without prior seizure history. No other acute morbidity was seen after radiosurgery.

Hemorrhage

Fifteen patients experienced intracerebral hemorrhages after radiosurgical treatment: (A) 0; (B) 0; (C) 2; (D) 13. Only five of these patients had a history of hemorrhage. Hemorrhages occurred 2, 3, 4, 4, 4, 5, 6, 6, 10, 11, 12, 13, 14, 26, and 34 months after radiosurgical treatment. Nine patients recovered fully, four had significant permanent neurologic deficits, and two died. Three of the patients who experienced hemorrhage were later documented to have angiographic cures (studies performed 23 and 50 months after radiosurgical treatment).

Radiation Edema/Necrosis

Eight patients (3%) experienced transient delayed complications directly attributable to radiosurgery. Two of these patients experienced headache, two had mild dysphasia, three had hemiparesis, and one had a field cut. The onset of symptoms occurred 7, 10, 10, 12, 14, 14, 14, and 15 months after radiosurgery. All had documented areas of edema adjacent to

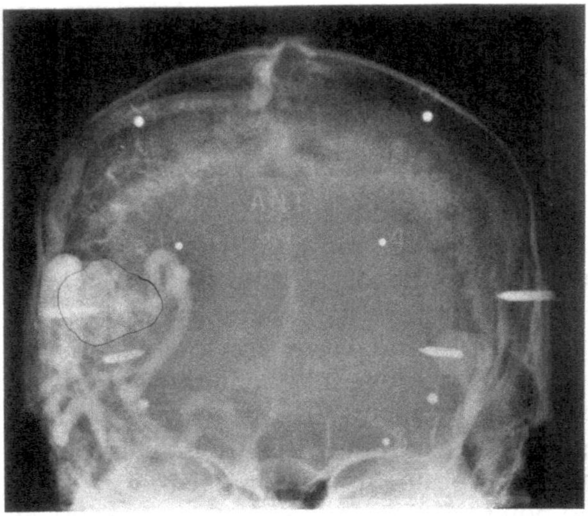

(A)
Part 1

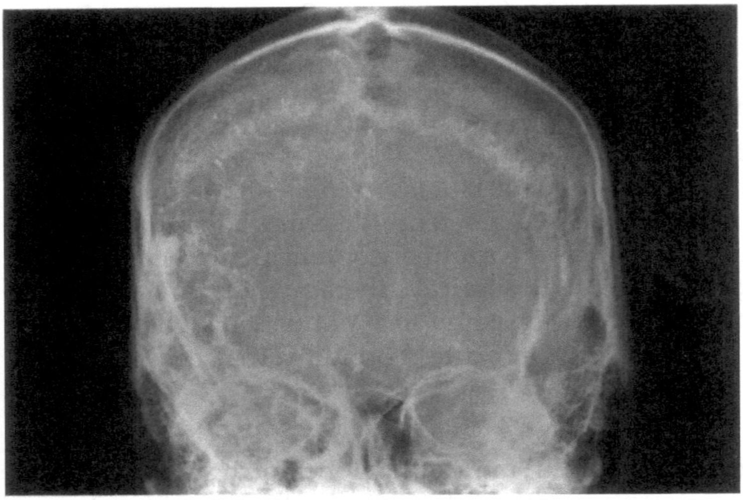

(B)
Part 1

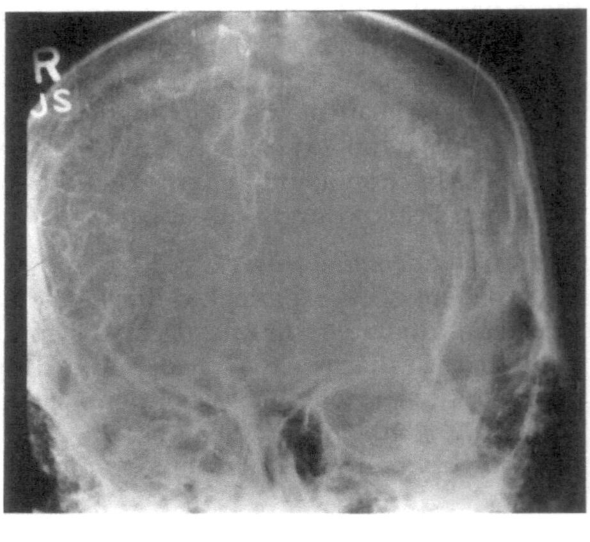

(C)
Par

Figure 10.2. (A) Original treatment-planning angiogram. Part 1: AP view; Part 2: lateral view. (B) Angiogram at 3 years showing decrease in size of resudual nidus. This is a definitive failure end point from the first radiosurgery treatment. Part 1: AP view; Part 2: lateral view. (C) Angiogram confirming ablation after retreatment of the AVM. Part 1: AP view; Part 2: lateral view.

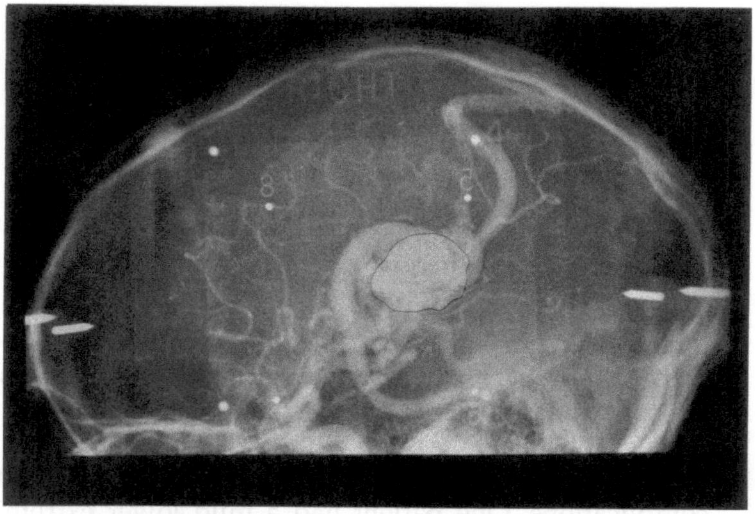

(A)
Part 2

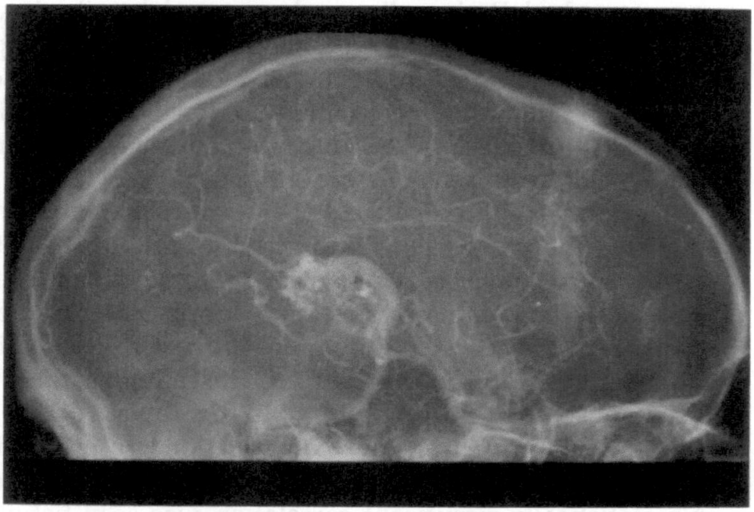

(B)
Part 2

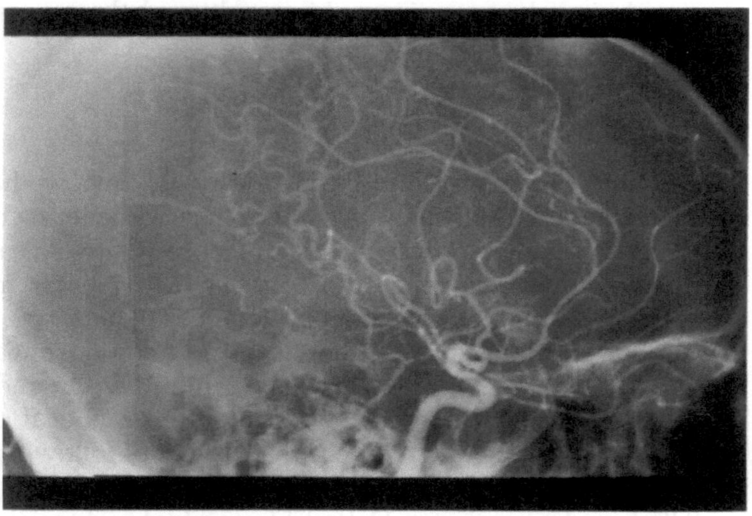

(C)
Part 2

Figure 10.2. *Continued*

Table 10.5. Definitive outcome analysis.

	A	B	C	D
Angiographic cure	1	29	22	17
Retreatment	1	3	3	18
Fatal hemorrhage	0	0	0	2
Percent success	50	91	88	46

their AVMs. In seven patients, the symptoms completely resolved; two were still under treatment with steroids. Four of these patients were subsequently documented to have angiographic cures.

Three patients (1%) experienced permanent radiation-induced complications. One patient had a mild lower extremity weakness, one had Parinaud's syndrome and hemibody analgesia, and one had a fourth nerve palsy. The onset of symptoms occurred 10, 11, and 14 months after radiosurgery. These patients were subsequently documented to have angiographic cures. Two of three patients had documented areas of edema, which resolved after months of steroid therapy.

In summary, three patients (1%) experienced minor, but permanent, neurologic deficits caused by radiation. Another eight patients (3%) experienced transient complications. The relationship between treatment dose and lesion size for the first 228 patients in the series is shown in Figure 7.4. Two patients with permanent complications received doses higher than what was used subsequently in other AVMs of similar volume, whereas the patients with transient complications received doses that have been safely used in other patients with similar-sized AVMs. No statistically significant correlation was seen between lesion size, lesion dose, angiographic result, and the occurrence of a complication.

Acoustic Schwannomas

At the University of Florida, as of May 21, 1996, we had treated 94 patients with acoustic schwannomas. Patients were selected for this treatment because of age (more than 65 years, 59 patients), tumor recurrence after surgery (23 patients), medical infirmity (2 patients), or patient preference (10 patients). The mean tumor volume was 5.9 cc, the mean radiation dose was 13.5 Gy, and the average follow-up was 32 months. On radiographic follow-up after treatment, two tumors grew. Local control for our first 56 patients with more than 1 year of follow-up revealed absolute local control of 98% and actuarial 5-year local control of 93% based on imaging. Delayed trigeminal and facial nerve palsies developed in 7 of these 56 patients (12.5%). Transient wors-

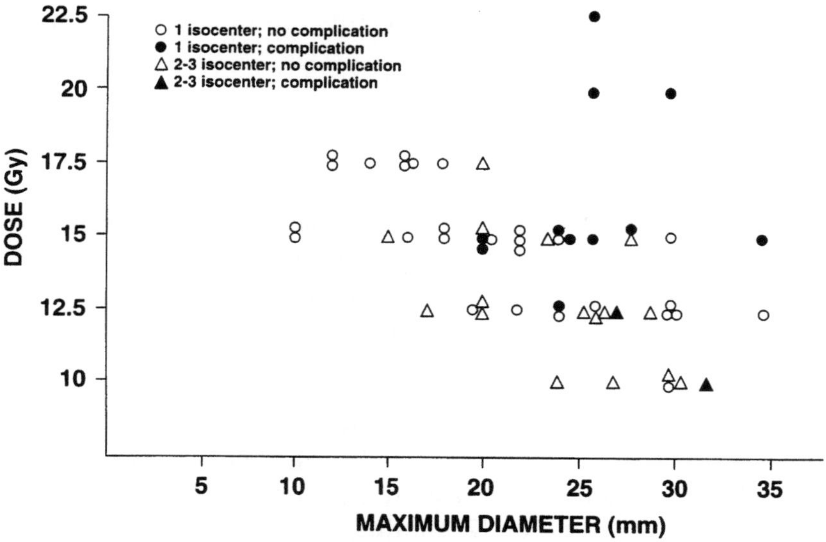

Figure 10.3. Scattergram of patients treated with acoustic schwannomas plotting diameter versus dose; patients with complications are indicated by solid points. The incidence of complications for lesions less than 24 mm treated to doses less than 15.0 Gy is low and has not occurred in patients treated with multiple-isocenter techniques.

ening of existing trigeminal or facial palsies occurred in an additional 5 patients. Three of these patients also required insertion of a shunt for symptomatic hydrocephalus. The fact that the risk of complications was related to dose and volume is worth noting. Patients in our series with a tumor <24 mm in diameter or receiving a prescribed dose <15 Gy were very unlikely to sustain even a transient complication (Fig. 10.3).

Meningiomas

Since 1988, 54 patients were treated at our institution for meningiomas. Mean tumor volume was 11.5 cc and mean tumor dose was 12 Gy (range, 10 to 20 Gy), usually prescribed to the 80% isodose shell. Thirty patients with 32 meningiomas had a minimum of 6 months of radiographic and clinical follow-up. The actuarial 2-year rate of local control in this group was 87%, with two tumors showing an increase in size. Both of these tumors had atypical histology. One patient with angioblastic meningioma had shrinkage of the treated lesion (i.e., local control), but died of distant metastases. Two severe complications occurred early in our experience when comparatively high doses of radiation were used. Prolonged steroid use was required in one patient, as was steroid use followed by resection of necrosis in a second patient. Preliminary results are in agreement with other published series.

Malignant Gliomas

Radiosurgery for the treatment of malignant gliomas is an investigational protocol at the University of Florida. From May 1989 through May 1996, a total of 50 patients were treated. Radiosurgery was given as adjuvant primary therapy after external beam treatment in some cases, but at the time of recurrence in others. Most patients in our experience received hyperfractionated accelerated external beam radiotherapy, which reduced the incidence of progressive disease before radiosurgery. Our initial 13 patients treated with this approach had a median actuarial survival time of 17 months, but all patients had progressive disease within 1 year of radiosurgery. In comparison with those of the University of Wisconsin and the Joint Center for Radiotherapy in Boston our results reveal that volume may be an important parameter in selecting patients for stereotactic radiosurgery boost. Furthermore, the decrease in progression observed after accelerated hyperfractionated treatment may bias our experience negatively because a significant number of patients in other published series are excluded from radiosurgery because of tumor progression or are patients that are made eligible for treatment because of decreased tumor volume after conventional treatment.

The combined experiences of our institution, Boston, and Wisconsin included 115 patients analyzed by the Radiation Therapy Oncology Group (RTOG) recursive partitioning technique to account for prognostic variables. The analysis found a statistically significant improvement in the 2-year survival rate in patients of intermediate prognosis. However, the technique did not fully account for the prognostic variables, and the Karnofsky Performance Status remained the best predictor of outcome.

Complications of stereotactic radiosurgery include prolonged steroid use and reoperation due to intractable edema. The incidence of these complications has ranged between 20 and 40% in individual series. These complications are similar to those observed with brachytherapy. A higher rate of complications is observed in patients with glioblastoma multiforme than with lower grade tumors, despite identical treatment doses.

We therefore continue to consider this treatment approach investigational and will further analyze the results as they become available. The RTOG opened a Phase III trial in 1994 of conventional radiotherapy with or without radiosurgery boost for glioblastoma multiforme.

Metastases

As of May 21, 1996, 107 patients with intracranial metastases had been treated with stereotactic radiosurgery at the University of Florida. Minimum eligibility requirements included histologic verification of primary disease, Karnofsky performance status >50%, four or fewer intracranial metastases, radiographically distinct lesion(s) <4 cm in diameter, and reasonably well-controlled primary disease.

A detailed analysis was performed of the first 70 patients treated: 34 patients were treated for progressive intracranial disease after external-beam radiotherapy, and 36 patients were treated with radiosurgery as an adjunct to initial treatment. Local control was achieved in 64% of lesions treated, with long-term follow-up on most patients and minimum follow-up of 6 months. Patients with lesions less than or equal to 24 mm in size had local control in 75% of cases. Transient symptomatology secondary to intracranial edema developed in eight patients (11%). All responded well to short courses of steroids. Univariate and multivariate analyses tested the prognostic significance of Karnofsky performance status, lesion volume, number of lesions, treatment dose, histology, site of primary disease, and time interval from primary diagnosis to development of intracranial metastases or from treatment of the metastasis to its recurrence. The only significant factor for survival after radiosurgery was lesion volume. Both lesion volume and treatment at presentation versus at recurrence were important in predicting local control with small lesion size and treatment at presentation being favorable characteristics. Patients treated with whole brain radiotherapy also had improved local control on univariate analysis.

Treatment of solitary metastases with radiosurgery has been difficult to analyze because of different selection criteria for surgery versus radiosurgery. A retrospective attempt to review patients treated with radiosurgery for solitary metastases who met all surgical resection criteria of Patchell showed radiosurgery was an effective approach to treatment. This review included patients treated at the University of Wisconsin, the Joint Center for Radiotherapy in Boston, the Medical College of Wisconsin, and our institution. The local control rate with radiosurgery exceeded 85%, and the only prognostic factors for survival in this group were the Karnofsky Performance Status and the presence of other metastases. This suggests that radiosurgery may be a preferable alternative for many patients with solitary metastases.

Conclusion

Our appreciation of the value of delivering focused high-dose radiation to radiographically well-defined targets has improved over time. It is clear that this experience along with the development of fractionated stereotactic techniques will enable even broader use of stereotactic radiation treatment approaches. Rigorous reporting of these results is the cornerstone of future development in this field.

11

Stereotactic Radiotherapy

Although-single fraction radiosurgery is of proven value for many types of intracranial disease, situations exist for which it cannot be used effectively. For example, consider an 8-year-old patient with residual craniopharyngioma after aggressive subtotal surgical resection. The residual tumor is densely adherent to the optic chiasm and hypothalamus. Single-fraction radiosurgery would therefore deliver the same dose of radiation to those structures as it would to the tumor capsule. A dose of >8 Gy has an unacceptable risk of injuring the optic structures and is clearly too low to achieve long-term tumor control. An alternative is to deliver a conventionally fractionated course of external radiotherapy. This treatment has proven efficacy, but it also has substantial risk of causing long-term cognitive deficits. The ideal treatment would incorporate conventional fractionation to spare the critical neurologic structures, but it would use stereotactic targeting and dosimetry techniques to reduce the volume of normal brain exposed to potentially harmful doses of radiation; hence, stereotactic radiotherapy (SRT).

The limiting factor for delivering SRT has historically been the need for rigid skull fixation. The application of a stereotactic head ring under local anesthesia is reviewed in Chapter 2. This procedure is well tolerated by most patients. However, very few patients would consent to daily application of the ring (e.g., 30 or more such treatments are needed for conventional fractionation). In addition, the repeated imaging and dosimetry procedures that would be required render this approach completely impractical. The solution to this problem must be noninvasive and reproducible. Two approaches have been studied: noninvasive stereotactic frames and frameless techniques.

Over the years, radiotherapists have developed a number of mask immobilization systems for use in conventional and stereotactic applications. In general, these techniques have an accuracy of 3 to 5 mm. Several noninvasive stereotactic frames have been developed. The Laitinen frame, which uses the ear canals and nasal bridge as fixation points, has been used extensively for both operative stereotaxis and radiosurgery. The Gill-Thomas-Cosman frame uses a bite plate and occipital rest to attach the patient to a BRW-style stereotactic head ring. This system has been tested extensively and used in radiosurgery. The median accuracy of both frames is in the 1 to 2 mm range, but the maximum inaccuracy may be much higher. Both systems require frame technology, albeit noninvasive, which is generally unfamiliar to the radiotherapy community. In addition, both systems use the same structure to perform localization and immobilization. Any stress that is placed on the noninvasive immobilization points (i.e., the bite plate or ear canals) may also affect the accuracy of localization.

At the University of Florida, an entirely frameless approach to SRT has been developed. A dental mold is fitted to the patient's upper teeth. This mold is coupled to an array of six infrared light-emitting diodes (LEDs) (Fig. 11.1). These LEDs are tracked by three infrared cameras mounted to the ceiling of the linac vault (Fig. 11.2). A computer program allows the position of the bite plate to be monitored in real time. Because the bite plate is attached to the upper jaw and the upper jaw is rigidly fixed to the remainder of the skull, this system allows any target in the brain to be located accurately, without the use of a frame.

On the first day of treatment, the normal radiosurgery paradigm is followed. After stereotactic frame application, imaging,

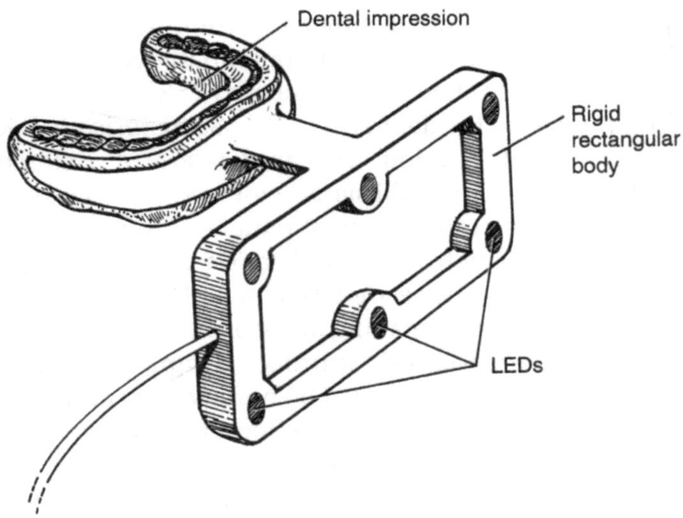

Figure 11.1. Bite plate system used for fractionated stereotactic radiotherapy. The infrared light-emitting diodes are attached to the rigid rectangular body that is coupled to the dental impression plate. Reproducibility of the plate for each patient is confirmed before use.

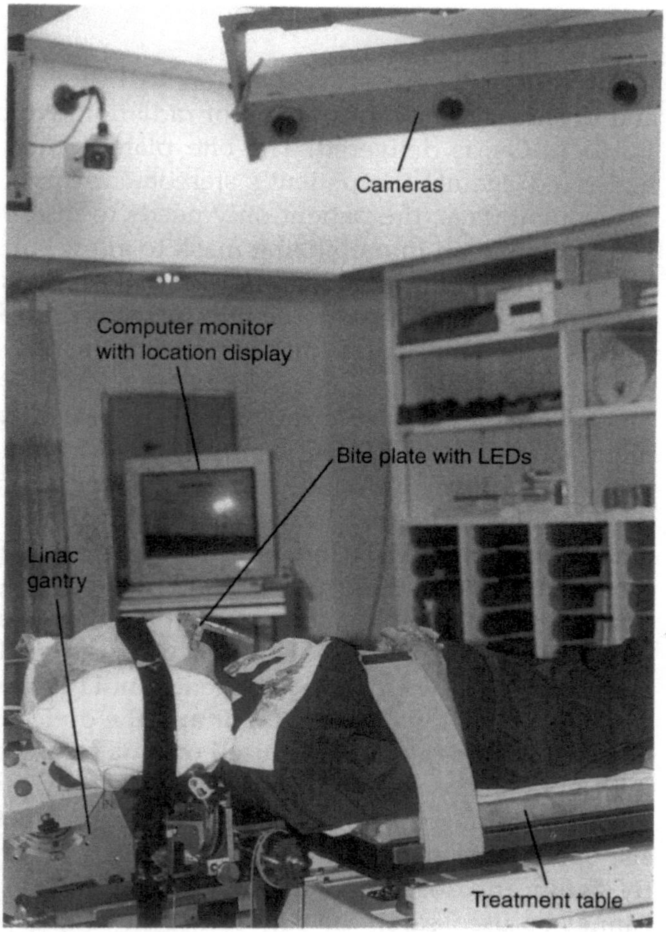

(A)

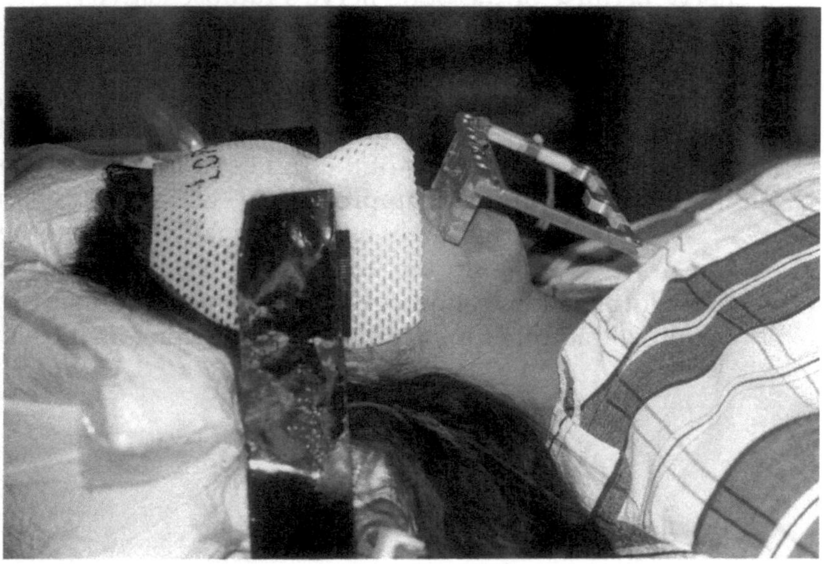

(B)

Figure 11.2. (A) The University of Florida fractionated stereotactic radiotherapy system showing the linac gantry, treatment table, camera system, and computer monitor that reports the location of the bite plate in real time. (B) The bite plate and immobilization system in use on a patient.

and dosimetry, the first fraction of radiation, which is typically 1.5 to 1.8 Gy, is delivered. The bite plate is then inserted and registered against the patient's stereotactic target position. On subsequent days, the patient only needs to insert the bite plate and lie within an immobilizing mask to allow subsequent repositioning for additional fractions of radiation. This system has a mean accuracy of 0.4 mm. The bite plate used for localization is *not* used for immobilization, thereby eliminating this potential source of error.

Forty-six patients have thus far been treated with this technique (see Table 11.1). For example, consider the case of a 50-year-old woman who had a 2-year history of gradual visual decline in her right eye. She was thoroughly evaluated by several neuroophthalmologists and, after MRI scanning (Fig. 11.3), was thought to have an optic nerve sheath meningioma. In general, it is not possible to remove these tumors without devascularizing the optic nerve, so the most common neurosurgical management approach is follow-up of such patients until they have lost useful vision, then removal of the tumor and the optic nerve. Conventional radiotherapy has been used on small numbers of patients with optic nerve meningiomas, with generally positive results. The disadvantage of conventional radiotherapy is that the opposite optic nerve and portions of the brain are invariably exposed to significant doses of radiation. Single-fraction radiosurgery cannot be used safely on any tumor in contact with the optic nerve because of the optic nerve's radiosensitivity. The patient had SRT in January 1995. She received a dose of 1.7 Gy per treatment via a 40-mm collimator (Fig. 11.3). A total of 30 fractions were administered using the bite-plate LED repeat positioning system. She tolerated the treatment well. Follow-up visual field examinations of the right eye in April 1995 surprisingly revealed dramatic improvement, which has persisted.

SRT is used increasingly as an alternative to radiotherapy in our practice. In the future, this type of easily applicable stereotactic localization technology may become a routine part of conventional radiotherapy for cranial and extracranial lesions.

Table 11.1. Patients treated with SRT.

	No. Pts.
Meningioma	13
Low grade glioma	9
Germinoma*	8
Aggressive pediatric*	5
Craniopharyngioma	4
Pineal*	2
Schwannoma	2
Chordoma*	1
Other	2

*Patients treated with combinations of conventional radiotherapy with part or all of radiation boost accomplished with fractionated stereotactic techniques.

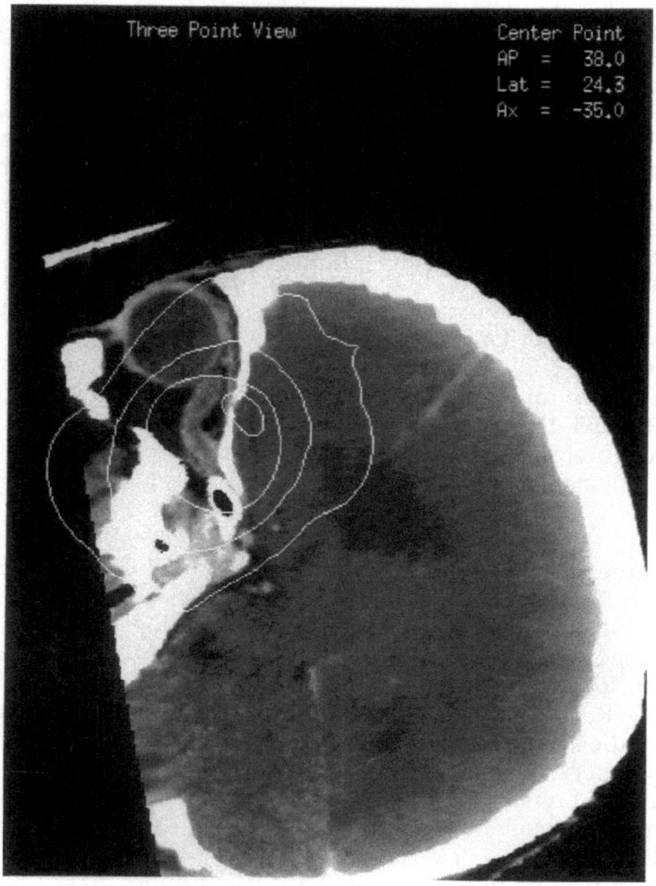

Figure 11.3. Optic nerve sheath meningioma treated with stereotactic radiotherapy. The isodose shells displayed are (beginning with the innermost) the 98%, 90%, 45%, and 18%. This distribution exposes significantly less normal tissue to high doses of radiation compared with conventional radiotherapy.

Suggested Readings

Acoustic Schwannomas

Barcia-Salorio JL, Hernandez G, Ciudad J, et al. Stereotactic radiosurgery in acoustic neurinoma. Acta Neurochir Suppl (Wien) 1984; 33:373–376.

Brackmann D, Kwartler JA. Treatment of acoustic tumors with radiotherapy. Arch Otolaryngol Head Neck Surg 1991; 116:161–162.

Delaney G, Matheson J, Smee R. Stereotactic radiosurgery: an alternative approach to the management of acoustic neuromas [letter]. Med J Austr 1992; 156:440.

Flickinger JC, Lunsford LD, Coffey RJ, et al. Radiosurgery of acoustic neurinomas. Cancer 1991; 67:345–353.

Flickinger JC, Lunsford LD, Linskey ME, et al. Gamma knife radiosurgery for acoustic tumors: multivariate analysis of four year results. Radiother Oncol 1993; 27:91–98.

Foote R, Coffey R, Swanson JR, et al. Stereotactic radiosurgery using the gamma knife for acoustic neuromas. Int J Radiat Oncol Biol Phys 1995; 32:1153–1160.

Hirsch A, Noren G. Audiologic findings after stereotactic radiosurgery in acoustic neurinomas. Acta Otolaryngol Suppl (Stockh) 1988; 449: 17.

Hirsch A, Noren G. Audiological findings after stereotactic radiosurgery in acoustic neurinomas. Acta Otolaryngol (Stockh) 1988; 106:244–251.

Hirsch A, Noren G, Anderson H. Audiologic findings after stereotactic radiosurgery in nine cases of acoustic neurinomas. Acta Otolaryngol (Stockh) 1979; 88:155–160.

Hudgins WR. Patients' attitude about outcomes and the role of gamma knife radiosurgery in the treatment of vestibular schwannomas. Neurosurgery 1994; 34:459–465.

Kamerer DB, Lunsford LD, Møller M. Gamma knife: an alternative treatment for acoustic neurinomas. Ann Otol Rhinol Laryngol 1988; 97:631–635.

Leksell L. A note on the treatment of acoustic tumors. Acta Chir Scand 1971; 137:763–765.

Linskey ME, Flickinger JC, Lunsford LD. Cranial nerve length predicts the risk of delayed facial and trigeminal neuropathies after acoustic tumor stereotactic radiosurgery. Int J Radiat Oncol Biol Phys 1993; 25:227–233.

Linskey ME, Lunsford LD, Flickinger JC. Neuroimaging of acoustic nerve sheath tumors after stereotaxic radiosurgery. Am J Neuroradiol 1991; 12:1165–1175.
Linskey ME, Lunsford LD, Flickinger JC. Radiosurgery for acoustic neurinomas: early experience. Neurosurgery 1990; 26:736–745.
Lunsford LD, Kamerer DB, Flickinger JC. Stereotactic radiosurgery for acoustic neuromas. Arch Otolaryngol Head Neck Surg 1990d; 116:907–909.
Maire JP, Floquet A, Darrouzet V, et al. Fractionated radiation therapy in the treatment of Stage III and IV cerebello-pontine angle neurinomas: preliminary results in 20 cases. Int J Radiat Oncol Biol Phys 1992; 23:147–152.
Mendenhall WM, Friedman WA, Bova FJ. Linear accelerator-based stereotactic radiosurgery for acoustic schwannomas. Int J Radiat Oncol Biol Phys 1994; 28:803–810.
Newman H, Sheline GE, Boldrey EB. Radiation therapy of tumors of the eighth nerve sheath. Am J Roentgenol Radium Ther Nucl Med 1974; 120:562–567.
Noren G, Arndt J, Hindmarsh T. Stereotactic radiosurgery in cases of acoustic neurinoma: further experiences. Neurosurgery 1983; 13:12–22.
Noren G, Arndt J, Hindmarsh T, et al. Stereotactic radiosurgical treatment of acoustic neurinomas. In: Lunsford LD, ed. Modern stereotactic neurosurgery. Boston: Martinus Nijhoff, 1988:481–490.
Noren G, Collins VP. Stereotactic biopsy in acoustic tumors. Appl Neurophysiol 1980; 43:189–197.
Noren G, Greitz D, Hirsch A, et al. Gamma knife surgery in acoustic tumours. Acta Neurochir (Wien) 1993; 58:104–107.
Ogunrinde OK, Lunsford LD, Flickinger JC, Kondziolka D. Stereotactic radiosurgery for acoustic nerve tumors in patients with useful preoperative hearing: results at 2 year followup examination. J Neurosurg 1994; 80:1011–1017.
Ogunrinde OK, Lunsford LD, Flickinger JC, et al. Cranial nerve preservation after stereotactic radiosurgery for small acoustic tumors. Arch Neurol 1995; 52:73–79.
To SY, Lufkin RB, Rand R, et al. Volume growth rate of acoustic neuromas on MRI post-stereotactic radiosurgery. Comput Med Imaging Graph 1990; 14:53–59.
Wallner KE, Sheline GE, Pitts LH, et al. Efficacy of irradiation for incompletely excised acoustic neurilemomas. J Neurosurg 1987; 6H7:858–863.

Complications and Central Nervous System Tolerance

Alexander E III, Siddon RL, Loeffler JS. The acute onset of nausea and vomiting following stereotactic radiosurgery: correlation with total dose to area postrema. Surg Neurol 1989; 32:40–44.
Anniko M. Early morphological changes following gamma irradiation. A comparison of human pituitary tumors and human acoustic neurinomas. Acta Pathol Microbiol Scand 1981; 89:113–124.
Flickinger JC. An integrated logistic formula for prediction of complications from radiosurgery. Int J Radiat Oncol Biol Phys 1989; 17:879–885.

Flickinger JC, Lunsford LD, Wu A, et al. Predicted dose-volume isoeffect curves for stereotactic radiosurgery with the ^{60}Co gamma unit. Acta Oncol 1991; 30:363–367.

Flickinger JC, Schell MC, Larson DA. Estimation of complications for linear accelerator radiosurgery with the integrated logistic formula. Int J Radiat Oncol Biol Phys 1990; 19:143–148.

Flickinger JC, Steiner L. Radiosurgery and the double logistic product formula. Radiother Oncol 1990; 17:229–237.

Hecht-Leavitt C, Grossman RI, Curran WJ, et al. MR of brain radiation injury: experimental studies in cats. Am J Neuroradiol 1987; 8:427–430.

Kjellberg RN, Koehler AM, Preston WM, Sweet WH. Intracranial lesions made by the Bragg peak of a proton beam. In: Haley TJ, Snider RS, eds. Response of the nervous system to ionizing irradiation. Boston: Little, Brown and Company, 1964:36–53.

Laitinen LV. Risks of radiosurgery [Letter]. Neurosurgery 1989; 25:480.

Leksell L, Herner T, Leksell D, et al. Visualisation of stereotactic radiolesions by nuclear magnetic resonance. J Neurol Neurosurg Psychiatry 1985; 48:19–20.

Leksell L, Larsson B, Andersson B, et al. Lesions in the depth of the brain produced by a beam of high energy protons. Acta Radiol 1960; 54:251–264.

Lindgren M. On tolerance of brain tissue and sensitivity of brain tumors to irradiation. Acta Radiol Suppl 1958; 170:1–73.

Loeffler JS, Siddon RL, Wen PY, et al. Stereotactic radiosurgery of the brain using a standard linear accelerator: a study of early and late effects. Radiother Oncol 1990; 17:311–321.

Marks JE, Wong J. The risk of cerebral radionecrosis in relation to dose, time and fractionation. A follow-up study. Prog Exp Tumor Res 1985; 29:210–218.

Marks LB, Spencer DP. The influence of volume on the tolerance of the brain to radiosurgery. J Neurosurg 1991; 75:177–180.

Mikhael MA. Radiation necrosis of the brain: correlation between patterns on computed tomography and dose of radiation. J Comput Assist Tomogr 1979; 3:241–249.

Nedzi LA, Kooy H, Alexander E III, et al. Variables associated with the development of complications from radiosurgery of intracranial tumors. Int J Radiat Oncol Biol Phys 1991; 21:591–599.

Nielsen SL, Kjellberg RN, Asbury AK, et al. Neuropathologic effects of proton-beam irradiation in man. I. Dose-response relationships after treatment of intracranial neoplasms. Acta Neuropathol (Berlin) 1972; 20:348–356.

Rizzoli HV, Paganelli DM. Treatment of delayed radiation necrosis of the brain: a clinical observation. J Neurosurg 1984; 60:589–594.

Schell MC, Smith V, Larson DA, et al. Evaluation of radiosurgery techniques with cumulative dose volume histograms in LINAC-based stereotactic external beam irradiation. Int J Radiat Oncol Biol Phys 1991; 20:1325–1330.

Shaw P, Bates D. Conservative treatment of delayed cerebral radiation necrosis. J Neurol Neurosurg Psychiatry 1984; 47:1338–1341.

Shearer DR, Roventine WA, Coy SR. Anomalous primary radiation from the Leksell Gamma Knife Unit. Health Phys 1994; 67:657–660.

Statham P, Macpherson P, Johnston R, et al. Cerebral radiation necrosis complicating stereotactic radiosurgery for arteriovenous malformation. J Neurol Neurosurg Psychiatry 1990; 53:476–479.
Thomsen J, Tos M, Borgesen SE. Gamma knife: hydrocephalus as a complication of stereotactic radiosurgical treatment of an acoustic neuroma. Am J Otol 1990; 11:330–333.
Tishler RB, Loeffler JS, Lunsford LD, et al. Tolerance of cranial nerves of the cavernous sinus to radiosurgery. Int J Radiat Oncol Biol Phys 1993; 27:215–221.
Tobias CA, Lawrence JH, Lyman JT, et al. Progress report on pituitary irradiation. In: Haley TJ, Snider RS, eds. Response of the nervous system to ionizing irradiation. Boston: Little, Brown and Company 1964:19–35.
Wennerstrand J, Ungerstedt U. Cerebral radiosurgery. II. An anatomical study of gamma radiolesions. Acta Chir Scand 1970; 136:133–137.
Woodruff KH, Lyman JT, Lawrence JH, et al. Delayed sequelae of pituitary irradiation. Hum Pathol 1984; 15:48–54.
Zeman W, Samorajski T. Effects of irradiation on the nervous system. In: Berdjis CC, ed. Pathology of irradiation. Baltimore: Williams & Wilkins 1971:213–277.

General Articles

Alexander E, Loeffler JS, Lunsford LD. Stereotactic radiosurgery. New York: McGraw-Hill, 1993.
Arndt J, Backlund EO, Larsson B, et al. Stereotactic irradiation of intracranial structures: physical and biological considerations. In: Szikla G, ed. Stereotactic cerebral irradiation (INSERM Symposium No. 12), Amsterdam: Elsevier 1979:81–92.
Backlund EO. Radiosurgery in intracranial tumors and vascular malformations. J Neurosurg Sci 1989; 33:91–93.
Barish RJ, Barish SV. A new stereotactic x-ray knife. Int J Radiat Oncol Biol Phys 1988; 14:1295–1298.
Barrow DL, Bakay RAE, Crocker I, et al. Stereotactic radiosurgery. J Med Assoc Ga 1990; 79:667–676.
Benassi M, Begnozzi L, Carpino S, et al. V. Magnetic resonance guided radiosurgery in children: tridimensional extrapolation from isodose neuroimaging superimposition. Childs Nerv Syst 1994; 10:115–121.
Betti OO, Derechinsky VE. Hyperselective encephalic irradiation with a linear accelerator. Acta Neurochir Suppl (Wien) 1984; 33:385–390.
Bradshaw JD. Special report. The stereotactic radiosurgery unit in Sheffield. Clin Radiol 1986; 37:277–279.
Burini G, Cassinari V, Giuliani G, et al. Theoretical study for use of Leksell's stereotactic frame in radiosurgery. J Neurosurg Sci 1989; 33:131–133.
Castro JR, Linstadt DE, Bahary JP, et al. Experience in charged particle irradiation of tumors of the skull base: 1977–1992. Int J Radiat Oncol Biol Phys 1994; 29:647–655.
Clark BG, Podgorsak EB, Souhami L, et al. A halo-ring technique for fractionated stereotactic radiotherapy. Br J Radiol 1993; 66:522–527.

Coffey RJ, Lunsford LD. Stereotactic radiosurgery using the 201 co-balt-60 source gamma knife. Neurosurg Clin N Am 1990; 1:933–954.

Colombo F. Linear accelerator radiosurgery. A clinical experience. J Neurosurg Sci 1989; 33:123–125.

Colombo F, Benedetti A, Casentini L, et al. Stereotactic radiosurgery of intracranial tumors in childhood. J Neurosurg Sci 1985; 29:233–237.

Colombo F, Benedetti A, Pozza F, et al. External stereotactic irradiation by linear accelerator. Neurosurgery 1985; 16:154–160.

Colombo F, Benedetti A, Pozza F, et al. Linear accelerator radiosurgery of three-dimensional irregular targets. Stereotact Funct Neurosurg 1990; 54–55:541–546.

Colombo F, Benedetti A, Pozza F, et al. Stereotactic radiosurgery utilizing a linear accelerator. Appl Neurophysiol 1985c; 48:133–145.

Colombo F, Benedetti A, Zanardo A, et al. New technique for three-dimensional linear accelerator radiosurgery. Acta Neurochir Suppl (Wien) 1987b; 39:38–40.

Colombo F, Casentini L, Pozza F, et al. Development of a second generation stereotactic apparatus for linear accelerator radiosurgery. Acta Neurochir Suppl (Wien) 1991; 52:84–86.

D'Agostino J, Pelczynski L. An overview of cyclotron treatment, Bragg peak proton hypophysectomy and Bragg peak radiosurgery for arteriovenous malformation of the brain. J Neurosurg Nurs 1979; 11:208–214.

Dahlin H, Sarby B. Destruction of small intracranial tumours with ^{60}Co gamma radiation. Acta Radiol 1975b; 14:209–227.

Delannes M, Daly NJ, Bonnet J, et al. Fractionated radiotherapy of small inoperable lesions of the brain using a non-invasive stereotactic frame. Int J Radiat Oncol Biol Phys 1991; 21:749–759.

Dempsey PK, Kondziolka D, Lunsford LD. Stereotactic diagnosis and treatment of pineal region tumours and vascular malformations. Acta Neurochir (Wien) 1992; 116:14–22.

Flickinger JC, Loeffler JS, Larson DA. Stereotactic radiosurgery for intracranial malignancies. Oncology (Huntingt) 1994; 8(1):81–97.

Friedman WA. LINAC radiosurgery. Neurosurg Clin N Am 1990; 1(4):991–1008.

Friedman WA, ed. Neurosurg Clin N Am, Philadelphia: WB Saunders, 1990; vol 1(4).

Friedman WA. Linear accelerator radiosurgery. In: DeSalles AF, Goetsch SJ, eds. Stereotactic surgery and radiosurgery. Madison, WI: Medical Physics Publishing, 1993:421–435.

Friedman WA, Bova FJ. Stereotactic radiosurgery. Contemp Neurosurg 1989; 11:1–7.

Friedman WA, Bova FJ. The University of Florida radiosurgery system. Surg Neurol 1989; 32:334–342.

Friedman WA, Bova FJ, Spiegelmann R. Linear accelerator radiosurgery at the University of Florida. Neurosurg Clin N Am 1992; 3:141–166.

Friedman WA, Spiegelmann R. Stereotactic surgery and radiosurgery. In: Little JR, Awad IA, eds. Reoperative neurosurgery. Baltimore: Williams & Wilkins, 1992:271–302.

Galloway RL, Maciunas RJ. Stereotactic neurosurgery. Crit Rev Biomed Eng 1990; 18:181–205.

Gill SS, Thomas DGT, Warrington AP, et al. Relocatable frame for stereotactic external beam radiotherapy. Int J Radiat Oncol Biol Phys 1991; 20:599–603.

Goodman ML. Gamma knife radiosurgery: current status and review. South Med J 1990; 83:551–554.

Graffman S, Brahme A, Larsson B. Proton radiotherapy with the Uppsala cyclotron. Experience and plans. Strahlenther Onkol 1985; 161:764–770.

Graham JD, Warrington AP, Gill SS, et al. A non-invasive, relocatable stereotactic frame for fractionated radiotherapy and multiple imaging. Radiother Oncol 1991; 21:60–62.

Greitz T, Lax I, Bergström M, et al. Stereotactic radiation therapy of intracranial lesions—methodologic aspects. Acta Radiol Oncol 1986; 25:81–89.

Hariz MI, Eriksson AT. Reproducibility of repeated mountings of a noninvasive CT/MRI stereoadapter. Appl Neurophysiol 1986; 49: 336–347.

Hariz MI, Henriksson R, Löfroth P-O, et al. A non-invasive method for fractionated stereotactic irradiation of brain tumors with linear accelerator. Radiother Oncol 1990; 17:57–72.

Hartmann GH, Schlegel W, Sturm V, et al. Cerebral radiation surgery using moving field irradiation at a linear accelerator facility. Int J Radiat Oncol Biol Phys 1985; 11:1185–1192.

Heifetz MD, Wexler M, Thompson R. Single-beam radiotherapy knife—a practical theoretical model. J Neurosurg 1984; 60:814–818.

Heifetz MD, Whiting J, Bernstein H, et al. Stereotactic radiosurgery for fractionated radiation: a proposal applicable to linear accelerator and proton beam programs. Stereotact Funct Neurosurg 1989; 53:167–177.

Hitchcock E, Kitchen G, Dalton E, et al. Stereotactic Linac radiosurgery. Br J Neurosurg 1989; 3:305–312.

Houdek PV, Fayos JV, Van Buren JM, et al. Stereotactic radiotherapy technique for small intracranial lesions. Med Phys 1985; 12:469–479.

Houdek PV, Schwade JG, Serago CF, et al. Computer controlled stereotaxic radiotherapy system. Int J Radiat Oncol Biol Phys 1991; 22:175–180.

Hudgins WR. What is radiosurgery? Neurosurgery 1988; 23:272–273.

Kirn TF. Proton radiotherapy: some perspectives. JAMA 1988; 259:787–788.

Kitchen G. Stereotactic radiosurgery [letter]. Clin Oncol 1989; 1: 118.

Kjellberg RN. Radiosurgery. Neurosurgery 1989; 25:670–672.

Kjellberg RN, Abbe M. Stereotactic Bragg peak proton beam therapy. In: Lunsford LD, ed. Modern stereotactic neurosurgery, Boston: Martinus Nijhoff, 1988:463–470.

Kjellberg RN, Koehler AM, Preston WM, et al. Stereotaxic instrument for use with the Bragg peak of a proton beam. Confin Neurol 1962; 22:183–189.

Kooy HM, Dunbar SF, Tarbell NJ, et al. Adaptation and verification of the relocatable Gill-Thomas-Cosman frame in stereotactic radiotherapy. Int J Radiat Oncol Biol Phys 1994; 30:685–691.

Kooy HM, van Herk M, Barnes PD, et al. Image fusion for stereotactic radiotherapy and radiosurgery treatment planning. Int J Radiat Oncol Biol Phys 1994; 28:1229–1234.

Kyuma Y, Hayashi A, Kitamura T, et al. Stereotactic radiosurgery using a linear accelerator. Neurol Med Chir (Tokyo) 1992; 32:572–577.

Larson DA, Bova F, Eisert D, et al. Current radiosurgery practice: results of an ASTRO survey. Int J Radiat Oncol Biol Phys 1993; 28:523–526.

Larson DA, Bova F, Eisert D, et al. Consensus statement on stereotactic radiosurgery quality improvement. Int J Radiat Oncol Biol Phys 1994; 28:527–530.

Larson DA, Gutin PH. Introduction to radiosurgery. Neurosurg Clin N Am 1990; 1:897–908.

Larson DA, Gutin PH, Leibel SA, et al. Stereotaxic irradiation of brain tumors. Cancer 1990; 65:792–799.

Larsson B. Biomedical program for the converted 200-MeV synchrocyclotron at the Gustaf Werner Institute. Radiat Res Suppl. 1985; 8:5310–5318.

Larsson B. On the application of a 185 MeV proton beam to experimental cancer therapy and neurosurgery: a biophysical study. Acta Univ Ups (Abstracts of Uppsala dissertations in medicine). 1962; 9:7–23.

Larsson B, Leksell L, Rexed B. The use of high energy protons for cerebral surgery in man. Acta Chir Scand 1963; 125:1–7.

Larsson B, Leksell L, Rexed B, et al. The high-energy proton beam as a neurosurgical tool. Nature 1958; 182:1222–1223.

Larsson B, Liden K, Sarby B. Irradiation of small structures through the intact skull. Acta Radiol 1974; 13:512–534.

Leksell L. Stereotactic radiosurgery. J Neurol Neurosurg Psychiatry 1983; 46:797–803.

Leksell L. The stereotaxic method and radiosurgery of the brain. Acta Chir Scand 1951; 102:316–319.

Leksell L. Stereotaxis and Radiosurgery. Springfield, IL: Charles C. Thomas, 1971.

Leksell D. Radiosurgery. Neurosurgery 1989; 24:297–298.

Leksell DG. Stereotactic radiosurgery. Present status and future trends. Neurol Res 1987; 9:60–68.

Levin CV, de Villiers JC, Jones DT. Radiosurgery for intracranial lesions. S Afr Med J 1990; 78:385–386.

Levy RP, Fabrikant JI, Frankel KA, et al. Charged-particle radiosurgery of the brain. Neurosurg Clin N Am, 1990; 1:955–990.

Lindquist C, Kihlstrom L, Hellstrand E. Functional neurosurgery—a future for the gamma knife? Stereotact Funct Neurosurg 1991; 57:72–81.

Loeffler JS, Alexander E III. The role of stereotactic radiosurgery in the management of intracranial tumors. Oncology 1990; 4:21–31.

Ludewigt BA, Chu WT, Phillips MH, et al. Accelerated helium-ion beams for radiotherapy and stereotactic radiosurgery. Med Phys 1991; 18:36–42.

Lunsford LD, Coffey RJ, Cojocaru T, et al. Image-guided stereotactic surgery: a 10-year evolutionary experience. Stereotact Funct Neurosurg 1990; 54–55:375–387.

Lunsford LD, Flickinger JC, Coffey RJ. Stereotactic gamma knife radiosurgery. Initial North American experience in 207 patients. Arch Neurol 1990; 47:169–175.
Lunsford LD, Flickinger JC, Lindner G, et al. Stereotactic radiosurgery of the brain using the first United States 201 cobalt-60 source gamma knife. Neurosurgery 1989; 24:151–159.
Lunsford LD, Maitz A, Lindner G. First United States 201 source cobalt-60 gamma unit for radiosurgery. Appl Neurophysiol 1987; 50:253–256.
Lutz W, Winston KR, Maleki N. A system for stereotactic radiosurgery with a linear accelerator. Int J Radiat Oncol Biol Phys 1988; 14:373–381.
Luxton G, Petrovich Z, Jozsef G, et al. Stereotactic radiosurgery: principles and comparison of treatment methods. Neurosurgery 1993; 32:241–259.
Lyman JT, Kanstein L, Yeater F, et al. A helium-ion beam for stereotactic radiosurgery of central nervous system disorders. Med Phys 1986; 13:695–699.
Lyman JT, Phillips MH, Frankel KA, et al. Stereotactic frame for neuroradiology and charged particle Bragg peak radiosurgery of intracranial disorders. Int J Radiat Oncol Biol Phys 1989; 16:1615–1621.
Maitz AH, Lunsford LD, Wu A, et al. Shielding requirements on-site loading and acceptance testing on the Leksell gamma knife. Int J Radiat Oncol Biol Phys 1990; 18:469–476.
Marks LB. Conventional fractionated radiation therapy vs. radiosurgery for selected benign intracranial lesions (arteriovenous malformations, pituitary adenomas, and acoustic neuromas). J Neurooncol 1993; 17:223–230.
McGinley PH, Butker EK, Crocker IR, et al. A patient rotator for stereotactic radiosurgery. Phys Med Biol 1990; 35:649–657.
McKenzie MR, Souhami L, Podgorsak EB, et al. Photon radiosurgery: a clinical review. Can J Neurol Sci 1992; 19:212–221.
Olivier A, de Lotbiniere A, Peters T, et al. Combined use of digital subtraction angiography and MRI for radiosurgery and stereoencephalography. Appl Neurophysiol 1987; 50:92–99.
Olivier A, Sadikot AF, Villemure J-G, et al. Fractionated stereotactic radiotherapy for intracranial neoplasms. Stereotact Funct Neurosurg 1992; 59:193–198.
Pastyr O, Hartmann GH, Schlegel W, et al. Stereotactically guided convergent beam irradiation with a linear accelerator: localization-technique. Acta Neurochir (Wien) 99:61–64.
Patil AA. Adaptation of linear accelerators to stereotactic systems. In: Lunsford LD, ed. Modern stereotactic neurosurgery. Boston: Martinus Nijhoff, 1988:471–480.
Patil AA. 1989. Radiosurgery with the linear accelerator. Neurosurgery 25:143.
Perry AM, Fox RA. Stereotactic radiosurgery. Australas Phys Eng Sci Med 1989; 12:233–240.
Peters TM, Clark J, Pike B, et al. Stereotactic surgical planning with magnetic resonance imaging, digital subtraction angiography and computed tomography. Appl Neurophysiol 1987; 50:33–38.
Peters TM, Clark JA, Pike GB, et al. Stereotactic neurosurgery planning on a personal-computer-based work station. J Digit Imaging 1989; 2:75–81.

Phillips MH, Frankel KA, Lyman JT, et al. Comparison of different radiation types and irradiation geometries in stereotactic radiosurgery. Int J Radiat Oncol Biol Phys 1990; 18:211–220.

Phillips MH, Stelzer KJ, Griffin TW, et al. Stereotactic radiosurgery: a review and comparison of methods. J Clin Oncol 1994; 12:1985–1999.

Podgorsak EB, Olivier A, Pla M, et al. Dynamic stereotactic radiosurgery. Int J Radiat Oncol Biol Phys 1988; 14:115–126.

Podgorsak EB, Pike GB, Olivier A, et al. Radiosurgery with high energy photon beams: a comparison among techniques. Int J Radiat Oncol Biol Phys 1989; 16:857–865.

Podgorsak EB, Pike GB, Pla M, et al. Radiosurgery with photon beams: physical aspects and adequacy of linear accelerators. Radiother Oncol 1990; 17:349–358.

Rosander K. Single-dose irradiation of the brain. Acta Univ Ups (Acta Universitatis Upsaliensis. Abstracts of Uppsala dissertations from the Faculty of Science). 1984; 734:5–36.

Rosenfeld JV, Thomas GW, Kaye AH. Stereotactic radiosurgery: does Australia need it? Aust N Z J Surg 1990; 60:499–501.

Samblas JM, Bustos JC, Gutierrez-Diaz JA, et al. Stereotactic radiosurgery of the foramen magnum region and upper neck lesions: technique modification. Neurol Res 1994; 16:81–82.

Serago CF, Lewin AA, Houdek PV, et al. Stereotactic target point verification of an X ray and CT localizer. Int J Radiat Oncol Biol Phys 1991; 20:517–523.

Shields CB, Guan T, Almond PR, et al. Radioneurosurgery using the LINAC scalpel: technique, indications, and literature review. J Ky Med Assoc 1993; 91:276–283.

Siddon RL, Barth NH. Stereotaxic localization of intracranial targets. Int J Radiat Oncol Biol Phys 1987; 13:1241–1246.

Souhami L, Olivier A, Podgorsak EB, et al. Fractionated stereotactic radiation therapy for intracranial tumors. Cancer 1991; 68:2101–2108.

Spiegelmann R, Friedman WA. Radiosurgery. Harefuah 1992; 123:119–122.

Stein BM, Mohr JP, Sisti MB. Is radiosurgery all that it appears to be? Arch Neurol 1991; 48:19–20.

Thompson BG, Coffey RJ, Flickinger JC, et al. Stereotactic radiosurgery of small intracranial tumors: neuropathological correlation in three patients. Surg Neurol 1990; 33:96–104.

Thomson ES, Gill SS, Doughty D. Stereotactic multiple arc radiotherapy. Br J Radiol 1990; 63:745–751.

Tsai J-S, Buck BA, Svensson GK, et al. Quality assurance in stereotactic radiosurgery using a standard linear accelerator. Int J Radiat Oncol Biol Phys 1991; 21:737–748.

Vandermeulen D, Suetens P, Gybels J, et al. Angiographic localizer ring for the BRW stereotactic system. Acta Neurochir Suppl (Wien) 1987; 39:15–17.

Walton L, Bomford CK, Ramsden D. The Sheffield stereotactic radiosurgery unit: physical characteristics and principles of operation. Br J Radiol 1987; 60:897–906.

Warrington AP, Laing RW, Brada M. Quality assurance in fractionated stereotactic radiotherapy. Radiother Oncol 1994; 30:239–246.

Winston KR, Lutz W. Linear accelerator as a neurosurgical tool for stereotactic radiosurgery. Neurosurgery 1988; 22:454–464.

Woo SY, Maor MH. Improving radiotherapy for brain tumors. Oncology (Huntingt) 1990; 4(6):41–45.
Zhang J, Levesque MF, Wilson CL, et al. Multimodality imaging of brain structures for stereotactic surgery. Radiology 1990; 75:435–441.

Malignant Gliomas

Chamberlain MC, Barba D, Kormanik P, et al. Stereotactic radiosurgery for recurrent gliomas. Cancer 1994; 74:1342–1347.
Coffey RJ. Boost gamma knife radiosurgery in the treatment of primary glial tumors. Stereotact Funct Neurosurg 1993; 61(Suppl 1):59–64.
Laing RW, Warrington AP, Graham J, et al. Efficacy and toxicity of fractionated stereotactic radiotherapy in the treatment of recurrent gliomas (phase I/II study). Radiother Oncol 1993; 27:22–29.
Laperriere NJ. Critical appraisal of experimental radiation modalities for malignant astrocytomas. Can J Neurol Sci 1990; 17:199–208.
Loeffler JS, Alexander E, Shea WM, et al. Radiosurgery as part of the initial management of patients with malignant gliomas. J Clin Oncol 1992; 10:1379–1385.
Mehta MP, Masciopinto J, Rozental J, et al. Stereotactic radiosurgery for glioblastoma multiforme: report of a prospective study evaluating prognostic factors and analyzing long-term survival advantage. Int J Radiat Oncol Biol Phys 1994; 30:541–549.
Nelson DF, Urtasun RC, Saunders WM, et al. Recent and current investigations of radiation therapy of malignant gliomas. Semin Oncol 1986; 13:46–55.
Sarkaria JN, Mehta MP, Loeffler JS, et al. Radiosurgery in the initial mamagement of malignant gliomas: survival comparison with the RTOG recursive partitioning analysis. Radiation Therapy Oncology Group. Int J Radiat Oncol Biol Phys 1995; 32:931–941.

Meningioma

Duma CM, Lunsford LD, Kondziolka D, et al. Stereotactic radiosurgery of cavernous sinus meningiomas as an addition or alternative to microsurgery. Neurosurgery 1993; 32:699–705.
Engenhart R, Kimmig BN, Hover KH, et al. Stereotactic single high dose radiation therapy of benign intracranial meningiomas. Int J Radiat Oncol Biol Phys 1990; 19:1021–1026.
Kondziolka D, Lunsford LD, Coffey RJ, et al. Stereotactic radiosurgery of meningiomas. J Neurosurg 1991; 74:552–559.
Lunsford LD. Contemporary management of meningiomas: radiation therapy as an adjuvant and radiosurgery as an alternative to surgical removal. J Neurosurg 1994; 80:187–190.
Sekhar LN, Jannetta PJ, Burkhart LE, et al. Meningiomas involving the clivus: a six-year experience with 41 patients. Neurosurgery 1990; 27:764–781.
Spiegelmann R, Friedman WA. The radiosurgical treatment of meningiomas. In: Schmidek HH, ed. Meningiomas and their surgical management. Philadelphia: WB Saunders, 1991:508–516.

Valentino V, Schinaia G, Raimondi AJ. The results of radiosurgical management of 72 middle fossa meningiomas. Acta Neurochir (Wien) 1993; 122:60–70.

Metastases

Adler JR, Cox RS, Kaplan I, et al. Stereotactic radiosurgical treatment of brain metastases. J Neurosurg 1992; 76:444–449.

Auchter RM, Lamond JP, Alexander E, et al. A multiinstitutional outcome and prognostic factor analysis of radiosurgery for resectable single brain metastasis. Int J Radiat Oncol Biol Phys 1996; 35:27–35.

Black P McL. Solitary brain metastases: radiation, resection, or radiosurgery? Chest 1993; 103:367S–369S.

Brada M, Laing R. Radiosurgery/stereotactic external beam radiotherapy for malignant brain tumours: the Royal Marsden Hospital experience. Recent Results Cancer Res 1994; 135:91–104.

Buatti JM, Friedman WA, Bova FJ, et al. Treatment selection factors for stereotactic radiosurgery of intracranial metastases. Int J Radiat Oncol Biol Phys 1995; 32:1161–1166.

Caron J-L, Souhami L, Podgordak EB. Dynamic stereotactic radiosurgery in the palliative treatment of cerebral metastatic tumors. J Neurooncol 1992; 12:173–179.

Coffey RJ, Flickinger JC, Bissonette DJ, et al. Radiosurgery for solitary brain metastases using the cobalt-60 gamma unit: methods and results in 24 patients. Int J Radiat Oncol Biol Phys 1991; 20:1287–1295.

Coffey RJ, Flickinger JC, Lunsford LD, et al. Solitary brain metastasis: radiosurgery in lieu of microsurgery in 32 patients. Acta Neurochir Suppl (Wien) 1991; 52:90–92.

Davey P, O'Brien P. Disposition of cerebral metastases from malignant melanoma: implications for radiosurgery. Neurosurgery 1991; 28:8–15.

Fuller BG, Kaplan ID, Adler J, et al. Stereotaxic radiosurgery for brain metastases: the importance of adjuvant whole brain irradiation. Int J Radiat Oncol Biol Phys 1992; 23:413–418.

Gutin PH, Wilson CB. Radiosurgery for malignant brain tumors. J Clin Oncol 1990; 8:571–573.

Kihlstrom L, Karlsson B, Lindquist C. 1992. Gamma knife surgery for cerebral metastasis. Implications for survival based on 16 years experience. Stereotact Funct Neurosurg 1993; 61(Suppl 1):45–50.

Kihlstrom L, Karlsson B, Lindquist Ch, et al. Gamma knife surgery for cerebral metastasis. Acta Neurochir Suppl (Wien) 1991; 52:87–89.

Lindquist C. Gamma knife surgery for recurrent solitary metastasis of a cerebral hypernephroma: case report. Neurosurgery 1989; 25:802–804.

Loeffler JS, Kooy HM, Wen PY, et al. The treatment of recurrent brain metastases with stereotactic radiosurgery. J Clin Oncol 1990; 8:576–582.

Mehta MP, Rozental JM, Levin AB, et al. Defining the role of radiosurgery in the management of brain metastases. Int J Radiat Oncol Biol Phys 1992; 24:619–625.

Somaza S, Kondqiolka D, Lunsford LD, et al. Stereotactic radiosurgery for cerebral metastatic melanoma. J Neurosurg 1993; 79:661–666.

Sturm V, Kimmig B, Engenhardt R, et al. Radiosurgical treatment of cerebral metastases. Method, indications and results. Stereotact Funct Neurosurg 1991; 57:7–10.
Sturm V, Kober B, Hover K-H, et al. Stereotactic percutaneous single dose irradiation of brain metastases with a linear accelerator. Int J Radiat Oncol Biol Phys 1987; 13:279–282.
Valentino V, Mirri MA, Schinaia G, et al. Linear accelerator and Greitz-Bergstrom's head fixation system in radiosurgery of single cerebral metastases. A report of 86 cases. Acta Neurochir 1993; 121:140–145.

Miscellaneous Sites of Disease and Indications

Austin-Seymour M, Munzenrider J, Goitein M, et al. Fractionated proton radiation therapy of chordoma and low-grade chondrosarcoma of the base of the skull. J Neurosurg 1989; 70:13–17.
Backlund EO. Studies on craniopharyngiomas. IV. Stereotaxic treatment with radiosurgery. Acta Chir Scand 1973; 139:344–351.
Backlund EO. Studies on craniophatyngiomas. III. Stereotaxic treatment with intracystic yttrium-90. Acta Chir Scand 1973; 139:237–247.
Backlund EO, Johansson L, Sarby B. Studies on craniopharyngiomas. II. treatment by stereotaxis and radiosurgery. Acta Chir Scand 1972; 138:749–759.
Backlund EO, Rahn T, Sarby B. Treatment of pinealomas by stereotaxic radiation surgery. Acta Radiol 1974; 13:368–376.
Backlund EO, Rahn T, Sarby B, et al. Closed stereotaxic hypophysectomy by means of 60 Co gamma radiation. Acta Radiol Ther Phys Biol 1972; 11:545–555.
Barcia-Salorio JL, Roldan P, Hernandez G, et al. Radiosurgical treatment of epilepsy. Appl Neurophysiol 1985; 48:400–403.
Barcia-Salorio JL, Roldan P, Lopez-Gomez L. Radiosurgery of central pain. Acta Neurochir Suppl (Wien) 1987; 39:159–162.
Buatti JM, Friedman WA, Bova FJ, et al. LINAC radiosurgery for locally recurrent nasopharyngeal carcinoma: rationale and technique. Head Neck 1995; 17:14–19.
Casentini L, Colombo F, Pozza F, et al. Combined radiosurgery and external radiotherapy of intracranial germinomas. Surg Neurol 1990; 34:79–86.
Chandler HC, Friedman WA. Radiosurgical treatment of a hemangioblastoma: case report. Neurosurgery 1994; 34:353–355.
Coffey RJ, Cascino TL, Shaw EG. Radiosurgical treatment of recurrent hemangiopericytomas of the meninges: preliminary results. J Neurosurg 1993; 78:903–908.
Heikkinen ER, Heikkinen MI, Sotaniemi K. Stereotactic radiotherapy instead of conventional epilepsy surgery. A case report. Acta Neurochir (Wien) 1992; 119:159–160.
Hellstrand E, Abraham-Fuchs K, Jernberg B, et al. MEG localization of interictal epileptic focal activity and concomitant stereotactic radiosurgery. A non-invasive approach for patients with focal epilepsy. Physiol Meas 1993; 14:131–136.
Kaplan ID, Adler JR, Hicks WL, et al. Radiosurgery for palliation of base of skull recurrences from head and neck cancers. Cancer 1992; 70:1980–1984.

Kondziolka D, Lunsford LD. Stereotactic radiosurgery for squamous cell carcinoma of the nasopharynx. Laryngoscope 1991; 101:519–522.

Kondziolka D, Lunsford LD, Flickinger JC. The role of radiosurgery in the management of chordoma and chondrosarcoma of the cranial base. Neurosurgery 1991; 29:38–46.

Leksell L. Cerebral radiosurgery. I. Gammathalanotomy in two cases of intractable pain. Acta Chir Scand 1968; 134:585–595.

Leksell L. Sterotaxic radiosurgery in trigeminal neuralgia. Acta Chir Scand 1971; 137:311–314.

Leksell L, Meyerson BA, Forster DM. Radiosurgical thalamotomy for intractable pain. Confin Neurol 1972; 34:264.

Mindus P, Bergstrom K, Levander SE, et al. Magnetic resonance images related to clinical outcome after psychosurgical intervention in severe anxiety disorder. J Neurol Neurosurg Psychiatry 1987; 50:1288–1293.

Munzenrider JE, Gragoudas ES, Seddon JM, et al. Conservative treatment of uveal melanoma: probability of eye retention after proton treatment. Int J Radiat Oncol Biol Phys 1988; 15:553–558.

Page KA, Wayson K, Steinberg GK, et al. Stereotaxic radiosurgical ablation: an alternative treatment for recurrent and multifocal hemangioblastomas. A report of four cases. Surg Neurol 1993; 40:424–428.

Pollock BE, Kondziolka D, Flickinger JC, et al. Preservation of cranial nerve function after radiosurgery for nonacoustic schwannomas. Neurosurgery 1993; 33:597–601.

Pozza F, Colombo F, Chierego G, et al. Low-grade astrocytomas: treatment with unconventionally fractionated external beam stereotactic radiation therapy. Radiology 1989; 171:565–569.

Rand RW, Jacques DB, Melbye RW, et al. Leksell Gamma Knife treatment of tic douloureux. Stereotact Funct Neurosurg 1993; 61(Suppl 1):93–102.

Rand RW, Khonsary A, Brown WJ, et al. Leksell stereotactic radiosurgery in the treatment of eye melanoma. Neurol Res 1987; 9:142–146.

Saunders WM, Char DH, Quivey JM, et al. Precision, high dose radiotherapy: helium ion treatment of uveal melanoma. Int J Radiat Oncol Biol Phys 1985; 11:227–233.

Steiner L, Forster DM, Leksell L, et al. Gammathalamotomy in intractable pain. Acta Neurochir (Wien) 1980; 52:173–184.

Physics and Dosimetry of Radiosurgery

Arcovito G, Piermattei A, D'Abramo G, et al. Dose measurements and calculations of small radiation fields for 9-MV x rays. Med Phys 1985; 12(6):779–784.

Bova FJ. Radiation physics. Neurosurg Clin N Am 1990; 1:4:909–931.

Chierego G, Marchetti C, Avanzo RC, et al. Dosimetric considerations on multiple arc stereotaxic radiotherapy. Radiother Oncol 1988; 12:141–152.

Dahlin H, Larsson B, Leksell L, et al. Influence of absorbed dose and field size on the geometry of the radiation-surgical brain lesion. Acta Radiol 1975; 14:139–144.

Flickinger JC, Lunsford LD, Wu A, et al. Treatment planning for gamma knife radiosurgery with multiple isocenters. Int J Radiat Oncol Biol Phys 1990; 18:1495–1501.

Flickinger JC, Maitz A, Kalend A, et al. Treatment volume shaping with selective beam blocking using the Leksell gamma unit. Int J Radiat Oncol Biol Phys 1990; 19:783–789.

Gehring MA, Mackie TR, Kubsad SS, et al. A three-dimensional volume visualization package applied to stereotactic radiosurgery treatment planning. Int J Radiat Oncol Biol Phys 1991; 21:491–500.

Goitein M. The inverse problem. Int J Radiat Oncol Biol Phys 1990; 18:489–497.

Guan TY, Almond PR, Park HC, et al. Imaging of radiation dose for stereotactic radiosurgery. Med Dosim 1993; 18:135–142.

Heifetz MD, Rosemark PJ, Wexler MC, et al. Rapid method for determination of isocenter of radiation gantry and alignment of laser beams for stereotactic radiosurgery. Stereotact Funct Neurosurg 1989; 53:46–48.

Houdek PV, VanBuren JM, Fayos JV. Dosimetry of small radiation fields for 10-MV x rays. Med Phys 1983; 10:333–336.

Kondziolka D, Dempsey PK, Lunsford LD, et al. A comparison between magnetic resonance imaging and computed tomography for stereotactic coordinate determination. Neurosurgery 1992; 30:402–407.

Kooy HM, Nedzi LA, Loeffler JS, et al. Treatment planning for stereotactic radiosurgery of intra-cranial lesions. Int J Radiat Oncol Biol Phys 1991; 21:683–693.

Kubsad SS, Mackie TR, Gehring MA, et al. Monte Carlo and convolution dosimetry for stereotactic radiosurgery. Int J Radiat Oncol Biol Phys 1990; 19:1027–1035.

Larsson B. Dosimetry and radiobiology of protons as applied to cancer therapy and neurosurgery. In: Thomas RH, Perez-Mendez V, eds. Advances in radiation protection and dosimetry in medicine. New York: Plenum, 1980:367–394.

Lax I. Target dose versus extratarget dose in stereotactic radiosurgery. Acta Oncol 1993; 32:453–457.

Leksell L, Lindquist C, Adler JR, et al. A new fixation device for the Leksell stereotaxic system. Technical note. J Neurosurg 1987; 66:626–629.

Marchetti C, Chierego G, Pozza F. A retrospective analysis concerning physical aspects and clinical dosimetry in radiosurgery. J Neurosurg Sci 1989; 33:127–130.

Nedzi LA, Kooy HM, Alexander E III, et al. Dynamic field shaping for stereotactic radiosurgery: a modeling study. Int J Radiat Oncol Biol Phys 1993; 25:859–869.

Pike B, Peters TM, Podgorsak E, et al. Stereotactic external beam calculations for radiosurgical treatment of brain lesions. Appl Neurophysiol 1987; 50:269–273.

Pike B, Podgorsak EB, Peters TM, et al. Dose distributions in dynamic stereotactic radiosurgery. Med Phys 1987; 14:780–789.

Pike GB, Podgorsak EB, Peters TM, et al. Three-dimensional isodose distributions in stereotactic radiosurgery. Stereotact Funct Neurosurg 1990a; 54–55:519–524.

Pike GB, Podgorsak EB, Peters TM, et al. Dose distributions in radiosurgery. Med Phys 1990; 17:296–304.

Podgorsak EB, Olivier A, Pla M., et al. Dynamic stereotactic radiosurgery. Int J Radiat Oncol Biol Phys 1988; 14:115–126.
Podgorsak EB, Olivier A, Pla M, et al. Physical aspects of dynamic stereotactic radiosurgery. Appl Neurophysiol 1987; 50:263–268.
Rice RK, Hansen JL, Svensson GH, et al. Measurements of dose distributions in small beams of 6 MV x-rays. Phys Med Biol 1987; 32:1087–1099.
Saw CB, Ayyangar K, Suntharalingam N. Coordinate transformations and calculation of the angular and depth parameters for a stereotactic system. Med Phys 1987; 14:1042–1044.
Schlegel W, Pastyr O, Bortfeld T, et al. Computer systems and mechanical tools for stereotactically guided conformation therapy with linear accelerators. Int J Radiat Oncol Biol Phys 1992; 24:781–787.
Serago CF, Houdek PV, Bauer-Kirpes B, et al. Stereotactic radiosurgery: dose-volume analysis of linear accelerator techniques. Med Phys 1992; 19:181–185.
Sixel KE, Podgorsak EB, Souhami L. Cylindrical dose distributions in pseudodynamic rotation radiosurgery: an experimental study. Med Phys 1993; 20:163–170.
Suh T-S, Bova F, Yoon SC, Shinn KS, et al. Optimisation of dose distribution for linear accelerator-based stereotactic radiosurgery. Med Biol Eng Comput 1993; 31:S23–S30.
Wu A, Lindner G, Maitz AH, Kalend AM, et al. Physics of gamma knife approach on convergent beams in stereotactic radiosurgery. Int J Radiat Oncol Biol Phys 1990; 18:941–949.
Yeung D, Palta J, Fontanesi J, et al. Systematic analysis of errors in target localization and treatment delivery in stereotactic radiosurgery (SRS). Int J Radiat Oncol Biol Phys 1993; 28:493–498.

Pituitary

Degerblad M, Rähn T, Bergstrand G, et al. Long-term results of stereotactic radiosurgery to the pituitary gland in Cushing's disease. Acta Endocrinol (Copenh) 1986; 112:310–314.
Kjellberg RN, Kliman B. Lifetime effectiveness—a system of therapy for pituitary adenomas, emphasizing Bragg peak proton hypophysectomy. In: Linfoot JA, ed. Recent advances in the diagnosis and treatment of pituitary tumors. New York: Raven Press, 1979:269–288.
Levy RP, Fabrikant JI, Frankel KA, et al. Heavy-charged-particle radiosurgery of the pituitary gland: clinical results of 840 patients. J Stereotact Funct Neurosurg 1991; 57:22–35.
Rahn T, Thoren M, Anniko M. Gamma irradiation effects on human ACTH-producing pituitary tumors in organ culture. Arch Otorhinolaryngol 1983; 238:209–215.
Rahn T, Thoren M, Hall K, et al. Stereotactic radiosurgery in Cushing's syndrome: acute radiation effects. Surg Neurol 1980; 14:85–92.
Thoren M, Rahn T, Guo WY, et al. Stereotactic radiosurgery with the cobalt-60 gamma unit in the treatment of growth hormone-producing pituitary tumors. Neurosurgery 1991; 29:663–668.
Thoren M, Rahn T, Hall K, et al. Treatment of pituitary dependent Cushing's syndrome with closed stereotactic radiosurgery by means of ^{60}Co gamma radiation. Acta Endocrinol 1978; 88:7–17.

Tobias CA. Pituitary radiation: radiation physics and biology. In: Linfoot JA, ed. Recent advances in the diagnosis and treatment of pituitary tumors. New York: Raven Press, 1979:221–243.
Witt TC, Kondziolka D, Flickinger JC, et al. Stereotactic radiosurgery for pituitary tumors. In: Kondziolka D, ed. Radiosurgery 1995. Basel (Switzerland): Karger, 1996 (vol 1): 55–65.

Radiobiology of Radiosurgery

Altschuler E, Lunsford LD, Kondziolka D, et al. Radiobiologic models for radiosurgery. Neurosurg Clin N Am 1992; 3:61–77.
Andersson B, Larsson B, Leksell L, et al. Histopathology of late local radiolesions in the goat brain. Acta Radiol 1970; 9:385–394.
Anniko M, Arndt J, Noren G. The human acoustic neurinoma in organ culture. II. Tissue changes after gamma irradiation. Acta Otolaryngol (Stockh) 1981; 91:223–235.
Anniko M, Arndt J, Rahn T, et al. Gamma irradiation effects on human growth hormone producing pituitary adenoma tissue. An analysis of morphology and hormone secretion in an in vitro model system. Acta Otolaryngol (Stockh) 1982; 93:485–500.
Bova FJ, Spiegelmann R, Friedman WA. A device for experimental radiosurgery. Stereotact Funct Neurosurg 1991; 56:213–219.
Brenner DJ, Martel MK, Hall EJ. Fractionated regimens for stereotactic radiotherapy of recurrent tumors in the brain. Int J Radiat Oncol Biol Phys 1992; 21:819–824.
Buatti JM, Friedman WA, Theele DP, et al. The lazaroid U74389G protects normal brain from stereotactic radiosurgery induced radiation injury. Int J Radiat Oncol Biol Phys 1996; 34:591–597.
Fike JR, Cann CE, Turowski K, et al. Radiation dose response of normal brain. Int J Radiat Oncol Biol Phys 1988; 14:63–70.
Friedman WA, Blatt DR, Bova FJ. Experimental radiosurgery. In: DeSalles AF, Goetsch SJ, eds. Stereotactic surgery and radiosurgery. Madison: Medical Physics Publishing, 1993:267–275.
Gaffey CT, Montoya VJ. Split-brain cats prepared by radiosurgery. Int J Radiat Biol Relat Stud Phys Chem Med 1973; 24:229–242.
Hall EJ, Brenner DJ. The radiobiology of radiosurgery: rationale for different treatment regimes for AVMs, malignancies. Int J Radiat Oncol Biol Phys 1993; 25:381–385.
Hamilton MG, Spetzler RF. The prospective application of a grading system for arteriovenous malformations. Neurosurgery 1994; 34:2–7.
Kamiryo T, Berk HW, Lee KS, et al. A stereotactic device for experimental gamma knife radiosurgery in rats. A technical note. Acta Neurochir (Wien) 1993; 125:156–160.
Kondziolka D, Lunsford LD, Claassen D, et al. Radiobiology of radiosurgery: Part 1. The normal rat brain model. Neurosurgery 1992; 31:271–279.
Kondziolka D, Lunsford LD, Claassen D, et al. Radiobiology of radiosurgery: Part II. The rat C6 glioma model. Neurosurgery 1992; 31:280–288.
Linskey ME, Martinez AJ, Kondziolka D, et al. The radiobiology of human acoustic schwannoma xenografts after stereotactic radiosurgery evaluated in the subrenal capsule of athymic mice. J Neurosurg 1993; 78:645–653.

Lo EH, Delapaz RL, Frankel KA, et al. MRI and PET of delayed heavy-ion radiation injury in the rabbit brain. Int J Radiat Oncol Biol Phys 1991; 20:689–696.

Lo EH, Fabrikant JI, Levy RP, et al. An experimental compartmental flow model for assessing the hemodynamic response of intracranial arteriovenous malformations to stereotactic radiosurgery. Neurosurgery 1991; 28:251– 259.

Lunsford LD, Altschuler EM, Flickinger JC, et al. In vivo biological effects of stereotactic radiosurgery: a primate model. Neurosurgery 1990; 27:373–382.

Nilsson A, Wennerstrand J, Leksell D, et al. Stereotactic gamma irradiation of basilar artery in cat. Preliminary experiences. Acta Radiol Oncol 1978; 17:150–160.

Parsons JT. Time-dose-volume relationships in radiation therapy. In: Million RR, Cassisi NJ, eds. Management of head and neck cancer: a multidisplinary approach, second edition, 1994:203–243.

Rexed B, Mair W, Sourander P, et al. Effect of high energy protons on the brain of the rabbit. Acta Radiol 1959; 53:289–299.

Rodriguez A, Levy RP, Fabrikant J. Experimental central nervous system injury after charged-particle irradiation. In: Gutin PH, Leibel SA, Sheline GE, eds. Radiation injury to the nervous system. New York: Raven Press, 1991:149–182.

Spiegelmann R, Friedman WA, Bova FJ, et al. LINAC radiosurgery: an animal model. J Neurosurg 1993; 78:638–644.

Vascular Lesions

Altschuler EM, Lunsford LD, Coffey RJ, et al. Gamma knife radiosurgery for intracranial arteriovenous malformations in childhood and adolescence. Pediatr Neurosci 1989; 15:53–61.

Barcia-Salorio JL, Hernandez G, Broseta J, et al. Radiosurgical treatment of carotid-cavernous fistula. Appl Neurophysiol 1982; 45:520–522.

Barcia-Salorio JL, Soler F, Hernandez G, et al. Radiosurgical treatment of low flow carotid-cavernous fistulae. Acta Neurochir Suppl (Wien) 1991; 52:93–95.

Barcia-Salorio JL, Barcia JA, Soler F, et al. Stereotactic radiotherapy plus radiosurgical boost in the treatment of large cerebral arteriovenous malformations. Acta Neurochir (Wien) 1993; 58:98–100.

Betti OO. Treatment of arteriovenous malformations with the linear accelerator. Appl Neurophysiol 1987; 50:262.

Betti OO, Munari C, Rosler R. Stereotactic radiosurgery with the linear accelerator: treatment of arteriovenous malformations. Neurosurgery 1989; 24:311–321.

Blatt DR, Friedman WA, Bova FJ. Modifications based on computed tomographical imaging in planning radiosurgical treatment of arteriovenous malformations. Neurosurgery 1993; 33:588–596.

Bova FJ, Friedman WA. Stereotactic angiography: an inadequate database for radiosurgery? Int J Radiat Oncol Biol Phys 1991; 20:891–895.

Chandler HC, Friedman WA. Successful radiosurgical treatment of a dural arteriovenous malformation. Neurosurgery 1993; 33:139–142.

Coffey RJ, Lunsford LD, Bissonette D, et al. Stereotactic gamma radiosurgery for intracranial vascular malformations and tumors:

report of the initial North American experience in 331 patients. Stereotact Funct Neurosurg 1990; 54–55:535–540.
Colombo F, Benedetti A, Casentini L, et al. Linear accelerator radiosurgery of arteriovenous malformations. Appl Neurophysiol 1987; 50:257–261.
Colombo F, Benedetti A, Pozza F, et al. Linear accelerator radiosurgery of cerebral arteriovenous malformations. Neurosurgery 1989; 24:833–840.
Colombo F, Pozza F, Chierego G, et al. Linear accelerator radiosurgery of cerebral arteriovenous malformations: an update. Neurosurgery 1994; 34:14–21.
Croft MJ. Stereotactic radiosurgery of arteriovenous malformations. Radiol Technol 1990; 61:375–379.
Crowell RM. Management of subarachnoid hemorrhage. Semin Neurol 1989; 9:210–217.
Dawson RC III, Tarr RW, Hecht ST, et al. Treatment of arteriovenous malformations of the brain with combined embolization and stereotactic radiosurgery: results after 1 and 2 years. Am J Neuroradiol 1990; 11:857–864.
Dias PS, Forster DMC. Management of cerebrovascular malformations. Br J Hosp Med 1989; 42:372–384.
Dias PS, Forster DM, Bergvall U. Cerebral medullary venous malformations. Report of four cases and review of the literature. Br J Neurosurg 1988; 2:7–21.
Duma ChM, Lunsford LD, Kondziolka D, et al. Radiosurgery for vascular malformations of the brain stem. Acta Neurochir (Wien) 1993; 58:92–97.
Engenhart R, Wowra B, Debus J, et al. The role of high-dose single fraction irradiation in small and large intracranial arteriovenous malformations. Int J Radiat Oncol Biol Phys 1994; 30:521–529.
Fabrikant JI, Lyman JT, Frankel KA. Heavy charged-particle Bragg peak radiosurgery for intracranial vascular disorders. Radiat Res Suppl 1985; 8:S244–S258.
Fabrikant JI, Lyman JT, Hosobuchi Y. Stereotactic heavy-ion Bragg peak radiosurgery for intra-cranial vascular disorders: method for treatment of deep arteriovenous malformations. Br J Radiol 1984; 57:479–490.
Fisher WS, Batjer HH, Friedman WA, et al. Surgery versus radiosurgery in the treatment of arteriovenous malformations. In: Hadley MN, ed. Perspectives in neurological surgery. St. Louis: Quality Medical Publishing, 1993:49–73.
Francel PC, Steiner L, Steiner M, et al. Repeat radiosurgical treatment in arteriovenous malformations following unsatisfactory results of initial single high-dose radiation. J Neurosurg 1991; 74:352A.
French LA, Chou SN. Conventional methods of treating intracranial arteriovenous malformations (Abstract). Prog Neurol Surg 1969; 3:274–319.
Friedman WA, Bova FJ. Linear accelerator radiosurgery for arteriovenous malformations. J Neurosurg 1992; 77:832–841.
Friedman WA, Bova FJ. Radiosurgery for ateriovenous malformations. Clin Neurosurg 1993; 40:446–464.
Friedman WA, Bova FJ, Mendenhall WM. Linear accelerator radiosurgery for arteriovenous malformations: the relationship of outcome to size. J Neurosurg 1995; 82:180–189.

Giorgi C, Cerchiari U, Broggi G, et al. 3-D reconstruction of cerebral angiography in stereotactic neurosurgery. Acta Neurochir Suppl (Wien) 1987; 39:13–14.

Gorzer H, Heimberger K, Schindler E. Spiral CT angiography with digital subtraction of extra- and intracranial vessels. J Comput Assist Tomogr 1994; 18:839–841.

Griffin BR, Warcola SH, Mayberg MR, et al. Stereotactic neutron radiosurgery for arteriovenous malformations of the brain. Med Dosim 1988; 13:179–182.

Guo WY, Karlsson B, Ericson K, Lindqvist M. Even the smallest remnant of an AVM constitutes a risk of further bleeding. Acta Neurochir (Wien) 1993; 121:212–215.

Guo W, Lindquist C, Karlsson B, et al. Gamma knife surgery of cerebral arteriovenous malformations: serial MR imaging studies after radiosurgery. Int J Radiat Oncol Biol Phys 1993; 25:315–323.

Guo WY, Lindqvist M, Lindquist C, et al. Stereotaxic angiography in gamma knife radiosurgery of intracranial arteriovenous malformations. Am J Neuroradiol 1992; 13:1107–1114.

Guo WY, Nordell B, Karlsson B, et al. Target delineation in radiosurgery for cerebral arteriovenous malformations. Acta Radiol 1993; 34:457–463.

Guo WY, Wikholm G, Karlsson B, et al. Combined embolization and gamma knife radiosurgery for cerebral arteriovenous malformations. Acta Radiol 1993; 34:600–606.

Halbach VV, Higashida RT, Yang P, et al. Preoperative balloon occlusion of arteriovenous malformations. Neurosurgery 1988; 22:301–308.

Heikkinen ER, Konnov B, Melnikov L, et al. Relief of epilepsy by radiosurgery of cerebral arteriovenous malformations. Stereotact Funct Neurosurg 1989; 53:157–166.

Heros RC, Korosue K, Diebold PM. Surgical excision of cerebral arteriovenous malformations: late results. Neurosurgery 1990; 26:570–578.

Hosobuchi Y, Fabrikant J, Lyman J. Stereotactic heavy-particle irradiation of intracranial arteriovenous malformations. Appl Neurophysiol 1987; 50:248–252.

Hudgins WR. Decision analysis of the treatment of AVMs with radiosurgery. Stereotact Funct Neurosurg 1993; 61(Suppl 1):11–19.

Kemeny AA, Dias PS, Forster DM. Results of stereotactic radiosurgery of arteriovenous malformations: an analysis of 52 cases. J Neurol Neurosurg Psychiatry 1989; 52:554–558.

Kjellberg RN. Stereotactic Bragg peak proton beam radiosurgery for cerebral arteriovenous malformations. Ann Clin Res 1986; 18(Suppl 47):17–19.

Kjellberg RN, Davis KR, Lyons S, et al. Bragg peak proton beam therapy for arteriovenous malformations of the brain. Clin Neurosurg 1983; 31:248–290.

Kjellberg RN, Hanamura T, Davis KR, et al. Bragg-peak proton-beam therapy for arteriovenous malformations of the brain. N Engl J Med 1983; 309:269–274.

Kondziolka D, Lunsford LD, Coffey RJ, et al. Stereotactic radiosurgery of angiographically occult vascular malformations: indications and preliminary experience. Neurosurgery 1990; 27:892–900.

Kondziolka D, Lunsford LD, Kanal E, et al. Stereotactic magnetic resonance angiography for targeting in arteriovenous malformation radiosurgery. Neurosurgery 1994; 35:585–591.

Laing RW, Childs J, Brada M. Failure of conventionally fractionated radiotherapy to decrease the risk of hemorrhage in inoperable arteriovenous malformations. Neurosurgery 1992; 30:872–876.

Levy RP, Fabrikant JI, Frankel KA, et al. Stereotactic heavy-charged-particle Bragg peak radiosurgery for the treatment of intracranial arteriovenous malformations in childhood and adolescence. Neurosurgery 1989; 24:841–852.

Lewis AI, Tomsick TA, Tew JM. Management of tentorial dural arteriovenous malformations: transarterial embolization combined with stereotactic radiation or surgery. J Neurosurg 1994; 81:851–859.

Lindquist C, Guo WY, Karlsson B, et al. Radiosurgery for venous angiomas. J Neurosurg 1993; 78:531–536.

Lindquist C, Steiner L. Stereotactic radiosurgical treatment of malformations of the brain. In: Lunsford LD, ed. Modern stereotactic neurosurgery. Boston: Martinus Nijhoff, 1988:491–506.

Lo EH. A theoretical analysis of hemodynamic and biomechanical alterations in intracranial AVMs after radiosurgery. Int J Radiat Oncol Biol Phys 1993; 27:353–361.

Loeffler JS, Alexander E III, Siddon RL, et al. Stereotactic radiosurgery for intracranial arteriovenous malformations using a standard linear accelerator. Int J Radiat Oncol Biol Phys 1989; 17:673–677.

Loeffler JS, Rossitch E Jr, Siddon R, et al. Role of stereotactic radiosurgery with a linear accelerator in treatment of intracranial arteriovenous malformations and tumors in children. Pediatrics 1990; 85:774–782.

Lunsford LD. Treatment of arteriovenous malformations by radiosurgery. In: Barrow DL, ed. Neurosurgical topics: Intracranial vascular malformations. Park Ridge: AANS, 1990:179–196.

Lunsford LD, Kondziolka D, Flickinger JC, et al. Stereotactic radiosurgery for arteriovenous malformations of the brain. J Neurosurg 1991; 75:512–524.

Marks MP, Delapaz RL, Fabrikant JI, et al. Intracranial vascular malformations: imaging of charged-particle radiosurgery. Part I. Results of therapy. Radiology 1988; 168:447–455.

Marks MP, Delapaz RL, Fabrikant JI, et al. Intracranial vascular malformations: imaging of charged-particle radiosurgery. Part II. Complications. Radiology 1988; 168:457–462.

Marks MP, Lane B, Steinberg GK, et al. Endovascular treatment of cerebral arteriovenous malformations following radiosurgery. AJNR Am J Neuroradiol 1993; 14:297–303.

Noorbehesht B, Fabrikant JI, Enzmann DR. Size determination of supratentorial arteriovenous malformations by MR, CT and angio. Neuroradiology 1987; 29:512–518.

Ogilvy CS. Radiation therapy for arteriovenous malformations: a review. Neurosurgery 1990; 26:725–735.

Ondra SL, Troupp H, George ED, et al. The natural history of symptomatic arteriovenous malformations of the brain: a 24-year follow-up assessment. J Neurosurg 1990; 73:387–391.

Paterson JH, McKissock W. A clinical survey of intracranial angiomas with special reference of their mode of progression and surgical treatment: a report of 110 cases. Brain 1991; 79:233–266.

Phillips MH, Frankel KA, Lyman JT, et al. Heavy charged-particle stereotactic radiosurgery: cerebral angiography and CT in the treatment of intracranial vascular malformations. Int J Radiat Oncol Biol Phys 1989; 17:419–426.

Phillips MH, Kessler M, Chuang FYS, et al. Image correlation of MRI and CT in treatment planning for radiosurgery of intracranial vascular malformations. Int J Radiat Oncol Biol Phys 1991; 20:881–889.

Pollock BE, Lunsford LD, Kondziolka D, et al. Patient outcomes after stereotactic radiosurgery for "operable" arteriovenous malformations. Neurosurgery 1994; 35:1–8.

Poulson MG. Arteriovenous malformations—a summary of 6 cases treated with radiation therapy. Int J Radiat Oncol Biol Phys 1987; 13:1553–1557.

Quisling RG, Peters KR, Friedman WA, et al. Persistent nidus blood flow in cerebral arteriovenous malformation after stereotactic radiosurgery: MR imaging assessment. Radiology 1991; 180:785–791.

Ray BS. Cerebral arteriovenous aneurysms. Surg Gynecol Obstet 1941; 73:615–648.

Redekop GJ, Elisevich KV, Gaspar LE, et al. Conventional radiation therapy of intracranial arteriovenous malformations: long-term results. J Neurosurg 1993; 78:413–422.

Sadler LR, Jungreis CA, Lunsford LD, et al. Angiographic technique to precede gamma knife radiosurgery for intracranial arteriovenous malformations. AJNR Am J Neuroradiol 1990; 11:1157–1161.

Saunders WM, Winston KR, Siddon RL, et al. Radiosurgery for arteriovenous malformations of the brain using a standard linear accelerator: rationale and technique. Int J Radiat Oncol Biol Phys 1988; 15:441–447.

Schad LR, Ehricke H-H, Wowra B, et al. Correction of spatial distortion in magnetic resonance angiography for radiosurgical treatment planning of cerebral arteriovenous malformations. Magn Reson Imaging 1992; 10:609–621.

Schwartz M, O'Brien P, Davey P, et al. Current status of radiosurgery for arteriovenous malformations. Can J Neurol Sci 1991; 18:499–502.

Seifert V, Stolke D, Mehdorn HM, et al. Clinical and radiological evaluation of long-term results of stereotactic proton beam radiosurgery in patients with cerebral arteriovenous malformations. J Neurosurg 1994; 81:683–689.

Sisti MB, Kader A, Stein BM. Microsurgery for 67 intracranial arteriovenous malformations less than 3 cm in diameter. J Neurosurg 1993; 79:653–660.

Smith HJ, Stother CM, Kikuchi Y, et al. MR imaging in the management of supratentorial intracranial AVMs. Am J Roentgenol 1988; 150: 1143–1153.

Souhami L, Olivier A, Podgorsak EB, et al. Dynamic stereotactic radiosurgery in arteriovenous malformation. Preliminary treatment results. Cancer 1990; 66:15–20.

Souhami L, Olivier A, Podgorsak EB, et al. Radiosurgery of cerebral arteriovenous malform with the dynamic stereotactic irradiation. Int J Radiat Oncol Biol Phys 1990; 19:775–782.

Spetzler RF, Martin NA. A proposed grading system of arteriovenous malformations. J Neurosurg 1986; 65:476–483.

Spiegelmann R, Friedman WA, Bova FJ. Limitations of angiographic target localization in planning radiosurgical treatment. Neurosurgery 1992; 30:619–624.
Steinberg GK, Fabrikant JI, Marks MP, et al. Stereotactic heavy-charged particle Bragg-peak radiation for intracranial arteriovenous malformations. N Engl J Med 1990; 323:96–101.
Steinberg GK, Levy RP, Fabrikant JI, et al. Stereotactic helium ion Bragg peak radiosurgery for angiographically occult intracranial vascular malformations. Stereotact Funct Neurosurg 1991; 57:64–71.
Steiner L. Treatment of arteriovenous malformations by radiosurgery. In: Wilson CB, Stein BM, eds. Intracranial arteriovenous malformations. Baltimore: Williams & Wilkins, 1984:295–313.
Steiner L. Radiosurgery in cerebral arteriovenous malformations. In: Fein JM, Flamm ES, eds. Cerebrovascular surgery, volume 4. New York: Springer-Verlag, 1985:1161–1215.
Steiner L, Leksell L, Forster DMC, et al. Stereotactic radiosurgery in intracranial arterio-venous malformations. Acta Neurochir Suppl (Wien) 1974; 21:195–209.
Steiner L, Leksell L, Greitz T, et al. Stereotaxic radiosurgery for cerebral arteriovenous malformations. Report of a case. Acta Chir Scand 1972; 138:459–464.
Steiner L, Lindquist C, Adler JR, et al. Clinical outcome of radiosurgery for cerebral arteriovenous malformations. J Neurosurg 1992; 77: 1–8.
Steiner L, Lindquist C, Cail W, et al. Microsurgery and radiosurgery in brain arteriovenous malformations. J Neurosurg 1993; 79:647–652.
Stelzer K, Griffin B, Eskridge J, et al. Results of neutron radiosurgery for inoperable arteriovenous malformations of the brain. Med Dosim 1992; 16:137–141.
Svien H, Peserico L. Regression in size of arteriovenous anomaly. J Neurosurg 1960; 17:493–496.
Tognetti F, Andreoli A, Cuscini A, et al. Successful management of an intracranial arteriovenous malformation by conventional irradiation. J Neurosurg 1985; 63:193–195.
Valentino V. Radiosurgery in cerebral tumours and AVM. Acta Neurochir Suppl (Wien) 1988; 42:193–197.
Wolkov HB, Bagshaw M. Conventional radiation therapy in the management of arteriovenous malformations of the central nervous system. Int J Radiat Oncol Biol Phys 1988; 15:1461–1464.
Wollin M, Kuruvilla A, Kagan AR, et al. Critique of "Stereotactic radiosurgery for intracranial arteriovenous malformations using a standard linear accelerator." Int J Radiat Oncol Biol Phys 1990; 18:1535–1536.
Yamamoto M, Jimbo M, Ide M, et al. Long-term follow-up of radiosurgically treated arteriovenous malformations in children: report of nine cases. Surg Neurol 1992; 38:95–100.
Yamamoto M, Jimbo M, Ide M, et al. Postradiation volume changes in gamma-unit treated cerebral arteriovenous malformations. Surg Neurol 1993; 40:485–490.
Yamamoto M, Jimbo M, Kobayashi M, et al. Long-term results of radiosurgery for arteriovenous malformation: neurodiagnostic imaging and histological studies of angiographically confirmed nidus obliteration. Surg Neurol 1992; 37:219–230.

Index